LIVING AND WORKING IN SPACE

SPACE EXPLORATION

LIVING AND WORKING IN SPACE

Ray Spangenburg
and 16153
Diane Moser

Facts On File
New York • Oxford • Sydney

It is difficult to say what is impossible, for the dream of yesterday is the hope of today and the reality of tomorrow.

—Robert Goddard

Living and Working in Space
Copyright ©1989 by Ray Spangenburg and Diane Moser
All rights reserved. No part of this book may be reproduced or utilized in any form or by any means, electronic or mechanical, including photocopying, recording, or by any information storage and retrieval systems, without permission in writing from the publisher. For information contact:

Facts On File, Inc.   Facts On File Limited
460 Park Avenue South   or   Collins Street
New York, New York 10016   Oxford OX4-1XJ
                                                         United Kingdom

Spangenburg, Ray, 1939-
        Living and working in space/by Ray Spangenburg and Diane Moser.
            p. cm. — (Space exploration; 2)
        Bibliography: p.
        Includes index.
        ISBN 0-8160-1849-9 (alk. paper): $22.95
        1. Space stations. 2. Space shuttles. I. Moser, Diane, 1944-
            II. Title. III. Series.
    TL797.S66 1989
    639.4—dc19   89-1213

British CIP data available on request

Facts On File books are available at special discounts when purchased in bulk quantities for businesses, associations, institutions, or sales promotion. Please contact the Special Sales Department at 212/683-2244. (Dial 1-800-322-8755, except in NY, AK, HI)

Text and Jacket Design by Ron Monteleone
Composition by Facts On File, Inc.
Printed in the United States of America

10 9 8 7 6 5 4 3 2 1

This book is printed on acid-free paper.

To all those
who have risked their lives in space—
and to those brave pioneers
who lost their lives
as they challenged this final frontier.

Virgil "Gus" Grissom	Francis Scobee
Edward White	Michael Smith
Roger Chaffee	Judith Resnik
Vladimir Komarov	Ellison Onizuka
Georgy Dobrovolsky	Ronald McNair
Vladislav Volkov	Gregory Jarvis
Viktor Patsayev	Christa McAuliffe

CONTENTS

PREFACE

Until the first manned space launches began in 1961, the adventure and challenge of living and working in space existed only in science fiction. Today it has become a daily reality (in fact, as we write this preface, Shuttle astronauts are flying a mission high overhead and cosmonauts are working aboard Mir). Space may well be humanity's final frontier, the last and ultimate region for exploration. Unquestionably, it challenges our ingenuity, our engineering expertise, our wisdom and our courage. Above all, it tests our spirit of survival.

Space Exploration is a series of four books that follows humankind's adventures in space. *Living and Working in Space* takes you aboard early Soviet and American space stations, the Salyut series and Skylab; it takes you on the Apollo-Soyuz Test Project, when Soviets and Americans shook hands in the sky; it takes you along on U.S. Shuttle flights and aboard the USSR's Mir. This book also explores the way satellites work for us in space and provides an overview of the many space programs, both national and international, now under way in other countries, including France, Japan, China, Canada, India and many others. Finally, *Living and Working in Space* takes you into the future, providing a glimpse of space adventures to come, including the U.S./International Space Station Freedom, planned for the late 1990s, as well as plans for Moon bases and a manned mission to Mars in the next century.

Other books in this series tell other parts of the saga with: the story of the first space scientists, test pilots, and early astronauts and cosmonauts, including the Apollo missions to the Moon (*Opening the Space Frontier*); a close look at the mysteries of the Solar System uncovered by planetary probes like the Voyager, VEGA and Viking missions (*Exploring the Reaches of the Solar System*); and close-ups of the lives of space pioneers from the beginning to the present (*Space People from A-Z*).

Together these books tell an exciting tale of human intelligence at its best—dreaming dreams, solving problems and achieving results. Today's world owes much to those who gave their work and their lives in the past and to those who venture today—both personally, in manned programs, and intellectually, through their work—into space.

ACKNOWLEDGMENTS

This book could never have happened without the help of countless individuals in both industry and government throughout the world. While we won't try to name them all, we appreciate the time so many took to provide photographs, drawings and information. A few stand out as extraordinary, and to them a special "thank you": Terry White, formerly of NASA's Johnson Space Center, for his memory of more than 25 years of American manned space exploration; Mike Gentry at JSC for his knack for locating photos; and Nick Johnson at Teledyne Brown Engineering for his generous contribution of drawings and information on the Soviet program. Also our thanks to four magazine editors who have given us a steady stream of fascinating assignments on space over the years: Tony Reichhardt of *Final Frontier*, John Rhea, formerly of *Space World*, and Kate McMains and Leonard David of *Ad Astra*. And for their supportive enthusiasm, our thanks to our agent Linda Allen and to James Warren and Deirdre Mullane, our editors at Facts On File. Without you all this book would not be.

INTRODUCTION

"*Poyekhali!* [Let's go!]" yelled the exuberant young man on that April day as the rockets beneath him surged, lifting him powerfully toward space. Yuri Gagarin, a 27-year-old Soviet Air Force pilot, had become the first human being ever to enter space, launched from the arid steppes of Central Asia. Not long before that, no one really even knew whether humans could survive in space, much less whether it was physically possible to send them there. The year was 1961.

Only three and a half years earlier, on October 4, 1957, an unimportant-looking aluminum sphere called "Sputnik" had forever changed the way human beings looked at their planet. About the size of a beach ball, Sputnik had been lobbed into space by the Soviet Union and orbited the Earth, winking its way through the night sky—visible and obvious proof that the "impossible" was possible. We could send a human-made satellite beyond the atmosphere and far enough beyond Earth's gravity that it would not fall back to the ground. Instead it circled around the planet high in the sky much like that natural satellite, the Moon.

Both Soviet and American engineers, technologists and physicists—as well as other visionaries in Great Britain and Europe—had been working on the concept for years, but most people had no idea before Sputnik that travel in space could be more than a "Buck Rogers" science-fiction idea.

By 1961, both the United States and the Soviet Union had successfully launched numerous satellites and unmanned spacecraft, leading up to the day humans would enter space. Then less than a month after Gagarin's historic flight, Alan Shepard made the first American manned spaceflight (though, Shepard's flight, unlike Gagarin's, was suborbital—not completing an orbit around the Earth). More manned flights, both Soviet and American, followed that year, and on February 20, 1962, John Glenn made the first American flight to orbit the Earth. Both countries had crossed the great threshold into space.

Today, literally hundreds of satellites from many nations circle the Earth. They collect data about our planet and how we use it, provide communication links between locations far from each other, and perform more critical and useful daily tasks for us than most people are even aware of. Both Americans and Soviets have built powerful space programs with an impressive history of accomplishments in the last half of the 20th century—including nine U.S. missions to the Moon (or around it) between 1968 and 1972. The Soviets have almost literally "moved" into space, with their nearly continuously manned space station Mir and long-duration missions lasting as long as a year. The U.S. Space Shuttle has soared regularly into space since 1981 (except for a two-and-a-half-year break following the Challenger disaster). From the Shuttle humans can assist with launching, retrieval and servicing of satellites. It also provides a laboratory in space for scientific and commercial experiments in zero gravity.

Who's to say what tomorrow may bring? The impossible has become possible; the uncertain has become real; the dream has come true. One thing is absolutely certain, however: The last half of this century has seen space become a place where human beings now both live and work—and we have begun to learn how to set it to work for us.

PART 1

LIVING ON A NEW FRONTIER

1

SALYUT: BUILDING A CABIN IN THE SKY

Nothing will stop us. The road to the stars is steep and dangerous. But we're not afraid.
—Yuri Gagarin

The Foundations

For almost as long as humans have dreamed of traveling in space, we have dreamed of living there, too, of building bases on the Moon and putting frontier settlements on the far-flung planets.

By early in the 20th century the concept of space stations—hovering out in space, unattached to any planet's surface—began to appear regularly in popular literature. Space pioneers Konstantin Tsiolkovsky in the Soviet Union and Hermann Oberth in Germany both contributed scientific theories in the 1920s that showed that we could escape Earth's gravity with rockets.

Finally, a stunning popular article published in 1952 by *Collier's* magazine really focused public at-tention in the United States on space travel as a real possibility. It was called "Crossing the Space Frontier," and its author was Wernher von Braun, father of the notorious German World War II V-2 rocket and adopted father of the American space program. Retitled, *Across the Space Frontier* came out in book form in 1952, the same year as Arthur C. Clarke's classic science fiction novel about space stations, *Islands in the Sky*.

Arthur C. Clarke also first proposed the concept of placing an artificial satellite in geosynchronous orbit [see box]. Today, the "Clarke orbit" is still the most common orbital position for communications, navigational and meteorological satellites.

These two concepts of the early 20th century—space stations and artificial satellites—by now have

---

What Keeps Things in Orbit?

Everyone knows that when you throw a rock up in the air it comes back down. So why don't satellites fall out of the sky? Well, they do—if they aren't launched high enough, in the right direction (or heading) and with enough speed (velocity).

Theories of motion and gravitation first developed by Sir Isaac Newton sometime between 1642 and 1727 explain how this works. Put simply, Newton said that (1) when something is motionless it tends to remain motionless and when something is moving it tends to keep moving; (2) the

3

change made by any force in the motion of an object depends upon the degree of force and the mass of the object; and (3) for every action there is an equal and opposite reaction.

These rules apply equally to the planets (satellites of the Sun), to our Moon (the Earth's natural satellite) and to artificial satellites (including space stations and even the Space Shuttle, which, in a way, are really also satellites). As it speeds through space, Earth tries to move in a straight line, but it is drawn into an orbit around the Sun by the Sun's gravitational pull. Even though the Sun is 93 million miles away, it still exerts this pull. The same, on a smaller scale, goes for the Moon and for satellites with respect to the Earth.

As a man-made satellite is launched by rockets, it streaks toward outer space. If it is not moving so fast that it escapes Earth's pull, it is caught in near space (close by Earth) and its direction is changed into an orbit around the planet. However, if it is moving fast enough to maintain some force in its original direction, it doesn't fall back to Earth. Together, the two forces (Earth's pull downward and the satellite's outward movement) keep it moving around and around.

Some orbits are so low, however, that Earth's atmosphere (although thin at these altitudes) enters into the picture, pulling (or dragging) on the satellite and slowing it down. When this happens, the orbit may "decay," that is, the satellite (or space station) loses the speed it needs to stay up and it falls back to Earth. Whether it breaks apart and burns, as it travels at high speeds through the atmosphere, or returns to Earth intact depends on whether it was designed to withstand the high temperatures and pressures of reentry.

Geosynchronous Orbit

Science fiction writer Arthur C. Clarke first popularized the idea of *geosynchronous orbit* in 1945. It's a special orbit 22,300 miles above the Earth, where the satellite's movement is synchronized (-*synchronous*) with the Earth (*geo-*). That is, a geosynchronous satellite orbits at the same speed that the Earth turns, once every 24 hours.

Geostationary Orbit

Seen from Earth, a geosynchronous satellite having a circular orbit directly over the equator seems to hover stationary in the sky—because it is in *geostationary orbit*. Because one geostationary satellite can reach 42 percent of the planet, this is an ideal orbit for communications satellites (often called "comsats"). As Clarke foresaw, by placing three of them spaced at 120 degrees apart it's possible to be in touch with almost the whole planet. Thanks to the hundreds of satellites now circling the planet in geostationary orbit, we have nearly complete global telecommunications today.

---

bloomed into a multi-billion-dollar, multi-national/commercial enterprise centering on the productive use of space. Now, as the new millennium approaches, we see artificial satellites taking on an ever increasing load of communications and Earth-viewing responsibilities, both civilian and military. In fact, many conveniences that we take for granted, from telephone communications to television broadcasts, depend on the now vast necklace of brilliant beads circling high above our planet.

We are in space to stay, and each day more and more of us make use of it. No longer is space the exclusive domain of the technological giants, the United States and the Soviet Union. Today space belongs to many nations—from Japan to West Germany, France to India, China to Sweden, and dozens of other countries, large and small, around the world. Tomorrow it will become the domain of many more. And all of us benefit from humanity's exploitation of its possibilities.

The artificial satellite will continue to change our world in many ways still beyond our imagination. In the near future space stations will do the same. In orbit only 20 to 60 minutes from Earth, a station permanently occupied by human crews on rotating shifts will be a cornerstone for space activities and space utilization ranging from Earth observation to astronomical studies, from satellite repair to materials processing. It can be a space laboratory, or a way station, a refueling point or a quarantine stop. A civilian "hotel" or a military "fort." Depending on the needs of its "guests," civilian or military, it can be any or all, or any combination.

With Freedom, the NASA/International Space Station, planned for the '90s, human beings will begin to occupy space on a permanent basis, not just as temporary visitors, but as residents, not just as trail blazers, but as colonists.

Even today, two small and primitive space stations are orbiting high overhead. Their names are Salyut (a "salute" to Yuri Gagarin, the first human in space) and Mir ("Peace"). It's not greatly surprising that both stations belong to the Soviet Union. As voiced by Konstantin Tsiolkovsky in 1903, stepping "out of the cradle" has long been a Soviet dream.

Getting There: Soyuz

During the early years after Sputnik began the space age, both Americans and Soviets developed their space programs along strikingly similar lines. While the Americans developed the Mercury and Gemini spacecraft in the mid-1960s that led up to the Apollo vehicle and landing on the Moon, the USSR produced first the simple Vostok ("East") spacecraft (1961-1963) and later its slightly expanded version, Voskhod ("Ascent") (1964-1965). Like Mercury and Gemini, the U.S. spacecraft of the early '60s, these first Soviet vehicles were extremely limited in sophistication and capability.

The development of the Apollo spacecraft, launched with its powerful Saturn rocket, made it possible for the Americans to go to the Moon. For the USSR, after Vostok and Voskhod, the future hinged on development of its workhorse spacecraft, Soyuz. From its first piloted flight in 1967 (which failed on reentry causing

the tragic death of its pilot, Vladimir Komarov), Soyuz had become a key element in the Soviet program. Able to carry a crew of one to three persons, Soyuz (its name means "Union") was capable of docking, as well as complex orbital maneuvers, rendezvous and extended flight lengths of up to 30 days. For the Soviets, the Soyuz technology provided the same kind of versatility and maneuverability that Apollo gave to the U.S.

Many observers believe that the Soyuz spacecraft was developed specifically to give the Soviets a try at the Moon. However, the USSR had begun planning for its first space station, Salyut, as early as 1967 and building a transportation system to the space station was clearly an important part of the Soyuz program, whether or not it might also have been designed to compete in a "race" against the U.S. to reach the Moon. In any case, by the time U.S. astronauts Neil Armstrong and Buzz Aldrin had walked on the Moon in 1969, the USSR had shifted full attention to building an outpost in the sky. With space stations in Earth orbit, a new era would begin, an age in which many of the Soviet dreams of space conquest would be attempted and achieved.

Early Frustration and Tragedy

In the beginning, Salyut was a small test station that allowed the USSR to study the effects of living in space for extended periods of time. Launched on April 19, 1971, Salyut was an important building-block in the Soviets' long-term space plans.

The station that would mark humankind's first attempt at putting a semi-permanent habitat in space was not an impressive vehicle. So small that it could be tucked today into the cargo bay of the U.S. Space Shuttle, Salyut 1 was primitive at best. It weighed about 20 tons and measured 65-1/2 feet long and 13 feet wide (20 m by 4 m). It was launched atop a D-class rocket—a medium-sized Proton launcher.

The crew of Soyuz 10 made the first trip. On April 23, 1971, lift-off went smoothly at the Soviet Tyuratam launch pad (called Baikonur cosmodrome by the Soviets, although the town of Baikonur is more than 200 miles away). Soyuz 10 ascended into orbit, rendezvoused with Salyut 1, and neared its target. With experienced cosmonaut Vladimir Shatalov on board as commander, docking was in capable hands as the Soyuz docking probe slipped neatly into place.

But, strangely enough, the crew never actually entered the station. After only five and a half hours joined to Salyut, Soyuz 10 backed off, took a tour around it to spot-check for damage, stayed aloft for another 16 hours and then headed home.

The reason for the shortened mission still remains mysterious. Was it some malfunction of the new hatch and docking tunnel system? Or possibly because Rukavishnikov, one of the cosmonauts, developed an especially severe bout of space sickness? In any case, it was a far cry from the 30-day mission they apparently had planned. However, the crew landed safely early in the morning of April 25 after a total mission time of just over two days.

The second crew, however, was not nearly so lucky. Everything started out smoothly enough. Launched June 6 aboard Soyuz 11, cosmonauts Georgy Dobrovolsky (commander), Vladislav Volkov and Viktor Patsayev docked with Salyut the next day. As the triumphant crew, led by Patsayev, opened the hatch of their docked Soyuz craft and floated through the tunnel, they became the first humans ever to inhabit a space station.

They had entered their somewhat cramped new quarters through one of Salyut's three pressurized modules, the access or transit module. The second—the main, center module—housed living and working quarters for the cosmonauts. A third module, for equipment, was also pressurized and housed the station's power supply, control panels, communications system and life-support systems. At the far end, the main engines and fuel were stowed aboard the unpressurized service module. Outside, four great solar panels extended like two parallel sets of wings to provide electrical power.

Setting up the station to make it livable took a full two days—since Salyut had been in mothballs ever since its launch seven weeks earlier. The crew fired up the Salyut engine to boost the decaying orbit, cavorted for Soviet TV viewers and put their orbiting laboratory in order.

A crowded schedule of scientific experiments faced the three during their stay—Earth observations (including studies of atmospheric processes, weather systems, snow and ice cover, and land and sea resources), astronomical studies and biological experiments with both animals and plants.

Many questions about life in space clamored for answers (and do still). Could humans grow their own food on a long space voyage? Would seeds sprout, grow and go to seed in weightlessness? No one knew the answers, so when seeds sprouted in Patsayev's greenhouse, called "Oasis," during the Soyuz 11 mission to Salyut the event seemed promising and the cosmonaut was elated.

The Soyuz 11 mission planners were aware that the human body was not designed for living in weightless space. Without gravity's pull, the human circulatory system tends to run amuck and muscles waste away because lifting and moving are practically effortless. To keep their muscles from becoming weak and useless, the crew worked out using basic exercise equipment. They wore special "penguin" exercise suits (so-called because of the penguin-like posture they cause) with elastic bands to create resistance, pulling against the cosmonauts' movements.

This was Volkov's second spaceflight. Patsayev celebrated his 38th birthday while hurtling around the Earth, as the three spent a historic 23 days aboard the station. It was an exciting pioneer mission with all the marks of triumph.

At the end of their stay, after loading their logs and materials back on Soyuz 11, the crew separated their craft from Salyut on June 29, settled into their seats and headed for home. Everything seemed to be going perfectly well.

On the ground, only the strange loss of communications with the crew less than an hour before landing hinted that all was not well. The automated landing

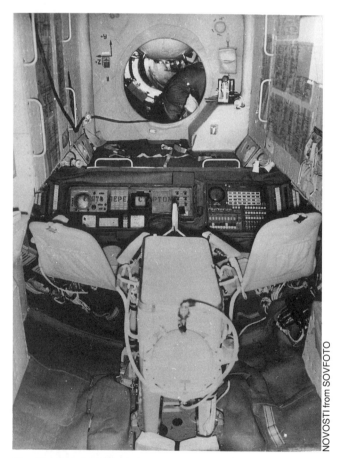

The flight deck of the Salyut 1 space station. Note hand holds for maneuvering the free fall and hatchway leading to Soyuz 11

NOVOSTI from SOVFOTO

went perfectly as the craft soft-landed on the Russian steppe. The recovery crew, arriving immediately on the scene via helicopter, hurriedly opened the hatch. They were completely unprepared for what they found: The three crew members were dead.

The planned celebrations turned to immense shock and grief for the Soviet people—in a reaction that extended beyond national boundaries, U.S. astronaut Tom Stafford served as one of the pallbearers. The three Soyuz 11 crew members were cremated, and their ashes, like Yuri Gagarin's and Vladimir Komarov's before them, were given a place of honor in the Kremlin Wall.

Dobrovolsky, Volkov and Patsayev were the first human beings to die in space, and their deaths raised many critical questions. Had we pushed beyond human ability to survive in space? The cosmonauts of a previous Soyuz mission—a stay of 17 days in space aboard Soyuz 9—had been so exhausted they had to be carried from their spacecraft on stretchers. Had the three Soyuz 11 cosmonauts now pushed space endurance too far? Had their bodies, unable to readjust quickly enough from weightlessness, been horribly crushed by Earth's gravity?

The answer, it turned out, was simpler than that. Investigation showed that during reentry a faulty valve was jarred open before the descent module had reached the Earth's atmosphere. Within a minute the air had been sucked out of the spacecraft, suddenly depressurizing the vehicle. The crew, traveling without space suits, and unprotected from the airless vacuum of space, had lost consciousness almost instantly, although we know from evidence found by the investigating team, that they apparently tried to reseal the valve. By the time the spacecraft reached the safety of the air below, it was much too late.

The tragedy stunned the Soviet program, but for several months Salyut 1 was kept ready for more visitors, with temperature and atmosphere maintained in life-support ranges. Another mission may have been planned. However, the Soyuz valve system couldn't be fixed in time. The Salyut orbit decayed, and the world's first space station self-destructed in the Earth's atmosphere on October 11, 1971. Its history brief and associated with frustration and tragedy, Salyut 1 had lasted only 175 days in orbit and only two crews of cosmonauts had reached its doorstep.

Try Again—and Again and Again

It was not a promising beginning for Salyut. The situation didn't get much better when an improved Salyut 2, launched on April 3, 1973, ran into trouble with its attitude control system and broke apart only

a few weeks later. Even though the station was unmanned, the loss was a major setback for the Soviet program.

Determined to outshine NASA's upcoming Skylab space station, the Soviets made another launch the following month. Unlike today's more open presentation of news by the USSR, news releases regarding space launches in the early '70s tended to be guarded. This launch was officially named Cosmos 557, and it was never designated as part of the Salyut series. However, when Western observers saw that this craft, too, was called down by ground control to a controlled destruction, they surmised it was actually, in all likelihood, yet another failed Salyut launch (a "second" Salyut 2).

In the meantime, the Soyuz 11 disaster had put a real damper on Soviet manned spaceflight. Two years passed before the next flight on September 27, 1973, when Oleg Makarov and physician Vasily Lazarev crewed the modified Soyuz craft.

In addition to new pressure equalization valves, the new Soyuz showed one other important change: Instead of three cosmonauts it now carried only two— and was refitted to accommodate space suits for reentry. Not until the Soyuz T-3 flight in 1980 would a Soviet craft fly again with more than a two-member crew.

Soyuz 12 tested out well in its short 47-hour, 16-minute stint, and set the stage for the Soyuz 13, which launched on December 18, 1973, for a 189-hour flight.

Like Makarov and Lazarev, crew Pyotr Klimuk and Valentin Lebedev also flew their missions without any Salyut space station in orbit to provide a destination. Coincidentally, the U.S. Skylab SL-4 mission was in progress, and for the first time Soviet and American crews orbited the Earth simultaneously. Although the Soviets had no communication with the American astronauts, both groups observed Comet Kahoutek, the cosmonauts using the Orion astrophysical observatory that was originally intended for use on a Salyut space station.

By June 25, 1974, the outlook became brighter yet with the successful launch of Salyut 3. Improved over earlier models, the four wing-like solar panels of Salyut 3 could now rotate toward the Sun for better power production. The new station also boasted better back-up systems and more comfortable accommodations for the crew.

Ten days after the Salyut launch, with the docking of Soyuz 14, its two-man crew breathed a sigh of relief. At last the first unmarred Soviet space station mission had begun. Living onboard in low-Earth orbit for just under two weeks (slightly over 377 hours), cosmonauts Pavel Popovich and Yuri Artyukhin ran through their medical and Earth-observation assignments smoothly and returned safely home.

A double-failure in the docking system forced an abort of the follow-up mission, Soyuz 15, launched the following month and abandoned after two days. Nonetheless, Salyut 3 had proved that the Soviets were getting back on course. Getting "on course" meant something a little more sinister, though, for Western observers who noted that the all-military crew of Soyuz 14 and the nature of some of its equipment and experiments indicated it was a military surveillance mission. Capable of semi-automatic operation, Salyut 3 telemetered data to ground receptors and automatically sent back to Earth at least one package—probably film—for Soviet retrieval teams to pick up. Military "spy missions" of this kind from space would prove increasingly important over the coming years, not only for the Soviets, but for the United States and other nations as well.

Salyut: A Series of Space Stations for the '70s and '80s

Basic configuration: Three inseparable, interdependent modules: Crew compartment (pressurized), docking transfer module (pressurized) and propulsion/instrumentation module (unpressurized). Up to five people can live and work aboard a Salyut at one time, with a usable space of about 3,531 cu. ft. (100 cu. m.). Solar panels supply the energy. Salyuts 1-5 had a single docking hatchway; a second was added with Salyut 6.
Size: 52.5 ft. long (16 m), 13.8 ft. wide (4.2 m)
Weight: About 19 tons

Salyut 1:

Solar panels arranged in three groups of two, with a total area of 452 sq. ft. (42 sq. m.).
Launch: April 19, 1971
Duration of Orbit: 175 days, to October 11, 1971
Missions Hosted: Soyuz 10-11
Records Set: First Soviet mission aboard a space station, lasting 23 days, 18 hours (Soyuz 11)
Problems: Possibly defective docking mechanism prevents cosmonauts from entering station (Soyuz 10). Accidentally depressurized Soyuz descent module causes death of three cosmonauts (Soyuz 11).

Salyut 2:

Launch: April 3, 1973
Duration of Orbit: 11 days, to April 14, 1973
Missions Hosted: None
Problems: The Salyut attitude control system becomes defective, the station begins to tumble and breaks apart.

Salyut 3:

Rotating solar panels add greater efficiency to this model of the Salyut station, along with improvements in back-up systems and comfort. However, Salyut 3 still has only one docking port.
Launch: June 25, 1974
Duration of Orbit: 7 months, to January 24, 1975.
Missions Hosted: Soyuz 14-15
Records Set: First successful Soviet space station mission (Soyuz 14).

Problems: Double failure in the Soyuz automatic docking system aborts mission (Soyuz 15).

Salyut 4:

Capable of carrying a wider range of scientific equipment, this "new improved model" could also be repaired and have parts replaced. Its solar panels (like those of the models to follow) were arranged in three groups, two horizontal and one vertical, each totalling an area of 215 sq. ft. (20 sq. m).

Launch: December 26, 1974
Duration of Orbit: 2 years, 40 days, to February 3, 1977
Missions Hosted: Soyuz 17, 18A (attempted) 18B, 20 (unmanned)
Records Set: First spacecraft to abort between lift-off and orbit (Soyuz 18A). Soviet endurance record of 62 days, 23 hours, 20 minutes (Soyuz 18B). "Crew" of tortoises and plants test the Salyut's self-sufficiency for an 89-day long- duration mission (Soyuz 20).
Problems: Two stages fail to separate at lift- off, aborting mission (Soyuz 18A).

Salyut 5:

Probably used primarily for military reconnaissance, such as high-resolution remote sensing.

Launch: June 22, 1976
Duration of Orbit: 1 year, 1 month, 16 days, (412 days) to August 8, 1977
Missions Hosted: Soyuz 21, 23-24
Records Set: Hook-up to a Soyuz spacecraft in a record 10 minutes (Soyuz 21). First non-Soviet instrument in a manned spacecraft (an East German-made camera carried aboard Soyuz 22—which did not, however, dock with Salyut).
Problems: Mission ended suddenly with reentry at night (Soyuz 21). Damaged Soyuz approach system causes docking failure and mission aborts (Soyuz 23). First Soviet splashdown—unintentional, at night (Soyuz 23).

Salyut 6:

The first Salyut designed for very long periods in orbit, Salyut 6 marks the beginning of a second generation of semi-permanent space stations. It had several new modifications, including: two docking hatches (forward, for crew entry, and rear, for resupply from the Progress cargo ferry), a new propulsion system (for repositioning the space station) that could be refueled in orbit, a water regeneration system (to purify and reuse water) and a shower for the cosmonauts. A redesigned space suit for space walks was also on board. New Delta onboard computer system introduced.

Launch: September 29, 1977
Duration of Orbit: 4 years, 10 months, to July 29, 1982
Missions Hosted: Soyuz 25 (attempted), 26-32, 33 (attempted), 34 (unmanned), 35-36, T-1 (unmanned), T-2, 37-38, T-3-4, 39-40
Major Accomplishments: Hosts a total of 16 Soyuz cosmonauts for a total of nearly 2 years (676 days) mission time in space. 1,600 scientific experiments carried out, including more than 800 biomedical experiments. 5,000 photographs of the Earth. Automatic docking with Cosmos 1267, prototype for future permanent stations.

Records Set: 96-day mission (Soyuz 26) breaks the 84-day record set in 1973/74 by the final U.S. Skylab mission (see chapter 2). Progress cargo ferry provides first unmanned resupply to a manned mission in space (launched January 20, 1978 for Soyuz 26). First Soyuz/Salyut crew to change vehicles in orbit. (Vladimir Dzhanibekov and Oleg Makarov, up on Soyuz 27, return on Soyuz 26.) First international crew; first non-Russian/non-American in space (with Czech pilot Vladimir Remek, Soyuz 28). New endurance record outpaces all others with 184 days, 19 hours, 2 minutes. (Leonid Popov and Valery Ryumin, Soyuz 35 to Salyut, Soyuz 37 return trip.) Record for total number of days in space for one person set by Ryumin, with 362 total days.

Problems: A mechanical failure in the Salyut docking mechanism prevents boarding (Soyuz 25).

Salyut 7:

Numerous new onboard features include: protective screens on portholes to avoid damage from micrometeorite impact; overall comfort improvements, such as new shower and refrigerator; improved space suits that permit space walks up to 5.5 hours. New equipment on board includes: Aelita, a biomedical diagnostic device connected to Salyut onboard processor for vascular and brain-related studies; a furnace for micro-gravity manufacturing; an X-ray telescope for astrophysical research; the gamma-ray telescope from Salyut 6.

Launch: April 19, 1982

Missions Hosted: Soyuz T-5 to T-9, T-10A (attempted), T-10B, T-11 to T-15.

Records Set: 211 days in space (set by Anatoly Berezovoy and Valentin Lebedev in Soyuz T-5). French pilot Jean-Loup Chrétien is first Westerner in orbit with a Soviet crew and spacecraft (Soyuz T-6). Svetlana Savitskaya (Soyuz T-7) becomes second woman in space, 19 years after Valentina Tereshkova. Leonid Kizim, Vladimir Solovyev and Oleg Atkov break endurance record with 236 days, 22 hours, 40 minutes (including transfer times from Earth via Soyuz T-10B to space station, February 8 to October 2, 1984). First working space walk by a woman, Svetlana Savitskaya (Soyuz T-12), July 25, 1984.

Problems: Mission aboard Salyut 7/Cosmos 1443 complex aborted due to problems with Soyuz T-8 rendezvous radar. Launch pad abort due to fire, crew lands safely (Soyuz T-10A).

The "New, Improved Models": Salyut 4-5

Salyut 3 would meet its end seven months after launch on January 24, 1975, but its replacement, Salyut 4, had already followed closely on its heels with a December 26, 1974, launch. This "new, improved model" could carry a wider range of scientific equipment and was designed so that cosmonauts could repair and replace parts in space. Unlike the earlier Salyut 1 to 3, Salyut 4 sported solar panels arrayed in three groups, two horizontal and one ver-

tical (instead of four horizontal "wings"). Each panel totaled an area of 215 square feet (20 sq. m).

Perhaps one of the most important innovations took place on the ground. At Tyuratam a new ground-based Salyut simulator—a replica of the real thing—had been built. In addition to its usefulness for training, ground crews could test out solutions to onboard problems with the new simulator and help advise mission crews about what to do in case of trouble.

During its two years, 40 days in orbit—more than three times longer than its predecessor—Salyut 4

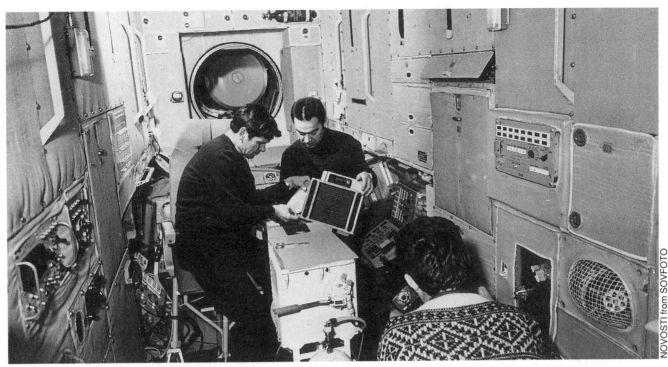

Alexei Gubarev and Georgy Grechko train in a Salyut 4 mockup

entertained three Soyuz missions (one unmanned). Aboard Soyuz 17, Alexei Gubarev and Georgy Grechko docked with Salyut 4 on January 13 and spent a trouble-free four weeks energetically performing scientific experiments in biology, physics and astronomy. They returned home 17 pounds lighter between them on February 9, 1975, having set a new Soviet record of 29 days in space. The station then saw Soyuz 18B cosmonauts Pyotr Klimuk and Vitaly Sevastyanov break that record with 62 days, 23 hours and 20 minutes in space, the world's second longest space mission at the time. In November 1975, Salyut 4 hosted a crew of tortoises and plants (Soyuz 20) that tested the unmanned station's self-sufficiency for an 89-day long-duration mission. The Soyuz 20 craft also tried out features of an automatically docking "cargo ferry," called Progress that the Soviets were developing. This slightly modified robotic Soyuz would run on automatic to bring supplies to the space station and was due to begin its runs two years later.

During the two-year life of Salyut 4 the only real problem occurred with the April 5, 1975, lift-off of Soyuz 18A, when two stages failed to separate. Achieving orbit generally requires more than one rocket, fired one after the other, a combination referred to as a "multistage rocket." After the first rocket (thought of as one "stage" in the process) has burned all its fuel, it drops away, leaving the next rocket, or stage, to take over. However, if the spent rocket doesn't separate due to a malfunction, the spacecraft cannot reach orbit, as happened in this case.

Soyuz 18A was the first spacecraft to abandon its mission between lift-off and orbit, but, thanks to the escape tower attached to the Soyuz craft, the crew members were jettisoned from the rocket. They plummeted toward Earth at a dizzying speed from their 90-mile height, and—although they worried that they would wind up in China, where they would have been less than welcome—they landed safely on the western slope of the Altai Mountains, just inside the Soviet Union near Mongolia.

Salyut 4's orbit deteriorated and it reentered Earth's atmosphere on February 3, 1977, according to plan and broke up over the Pacific Ocean, its job finished. Meanwhile, seven months earlier Salyut 5 had been successfully launched on June 22, 1976. Apparently used primarily for military reconnaissance, like Salyut 3, Salyut 5 hosted three Soyuz missions— Soyuz 21, 23 and 24. Much of their activity involved extensive photography and other forms of remote sensing, such as spectography. (Soyuz 22, launched on September 15, carried cosmonauts Colonel Valery Bykovsky [commander] and Vladimir Aksyonov. The craft orbited at an inclination that made docking with Salyut impossible—and no docking was probably

NOVOSTI from SOVFOTO

Drawing of Soyuz 18 undocking from Salyut 4

NOVOSTI from SOVFOTO

ever planned. Their mission—carrying a special multi-spectral camera built in East Germany—was most likely military.)

Although the Soyuz 21 mission was originally planned to last two months, the crew came home three weeks early, in August 1976, apparently because of atmosphere contamination aboard the station. Two months later, with the problem evidently cleaned up, Soyuz 23 was knocking on the door. Trouble with the Soyuz guidance system made docking impossible once again, however, and the crew headed for home, making an emergency landing in darkness.

In February 1977 the Soyuz 24 crew swooped in for one last mission to finish up the experiments begun by the Soyuz 21 pair the summer before. In the end, the results were impressive—the cosmonauts completed more than 300 experiments on board and photographed some 40 million square miles (65 million sq. km) of the Earth's surface, focusing on the Indian and Atlantic Ocean basins and the vast expanses of the Soviet Union.

Ten years had passed since the first Soyuz flight crashed to Earth on reentry, in the tragedy of 1967, and six since the disaster of Soyuz 11 when a malfunctioning valve left three cosmonauts lifeless as they returned to Earth. Through determination and persistence, the Soviets had learned from their failures and were beginning to replace those black memories with a new track record of success. Ahead lay ever-broadening experience in human spaceflight and ever-increasing expertise as they continued to rebuild new outposts in the sky.

2

THE BIRD WITH THE BROKEN WING: THE SKYLAB STORY

*M*obility around here is super...Every kid in the
United States would have a blast up here!
—Pete Conrad

In the United States, the Apollo program, with its first test flight in 1968 and missions to the Moon from 1969 to 1972, had been a huge success. A long-held dream had been fulfilled: Humans had walked on the Moon. By 1973 the American space program was at a peak and for many space enthusiasts there was nowhere to go now but up—and outward! They saw Apollo as the base from which all dreams would flow—from a manned lunar station to manned stations in geosynchronous and low-Earth orbit, and eventually manned planetary missions. Skylab was to be a small part of those grand dreams—an American space station, larger than Salyut, that could provide home and work space for a crew of three.

Aside from space visionaries, however, budget-cutters in Congress and the public in general didn't seem to catch the spirit of these ambitious dreams. The United States had won the race to the Moon and the social and political issues of the 1970s—such as the civil rights and peace movements—had begun to take top billing instead. With the war in Vietnam taking even larger and larger bites out of the United States economy, something had to give. There was no place for a space program in a society engaged in an escalating war overseas and beset by social problems at home. By the time the cost-cutting was over all that was left of the post-Apollo dream of lunar bases and

missions to Mars in the 20th century was an orphan called "Skylab" and a hybrid called the "Shuttle." Yet these two projects would provide the basis for the United States future in space.

Originally conceived as part of the Apollo Extension Program (later called the Apollo Applications Program), Skylab used surplus Apollo hardware—Saturn booster rockets and two segments of the Apollo spacecraft, the command module and the service module (often referred to together as the CSM). It was a concept filled with "can-do" spirit, economical, efficient and adventurous.

Originally, the whole workshop was going to be built in orbit, inside its own rocket fuel tank. In this "wet workshop" concept, the idea was to use the hydrogen tank and its fuel to get into space, then completely empty whatever fuel remained (hence the term "wet"), pump in a life-supporting atmosphere, fill the tank with equipment in space and set up a kind of on-the-spot science workshop.

It was an ingenious and colorful plan. However, when Congress decided to curtail the Apollo program, canceling several planned missions to the Moon, a powerful Saturn 5 launch vehicle (which would have been used to go to the Moon) became available for Skylab and the "wet workshop" gave way to a "dry"

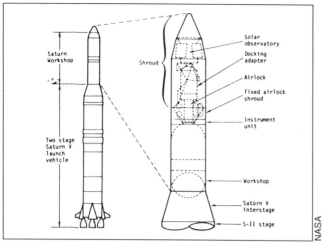

The Skylab workshop, constructed from the shell of an empty Saturn 4-B rocket tank, was launched atop a giant two-stage Saturn 5 rocket

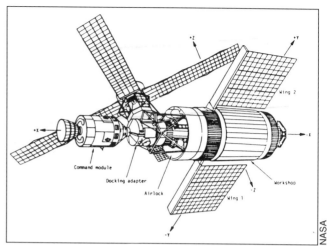

Skylab in orbit, with Apollo command/service module docked

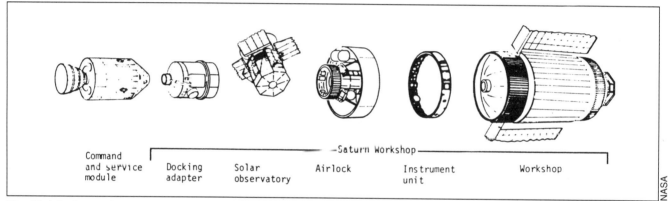

The Skylab cluster elements

one, assembled and outfitted on Earth. Saturn 5 was strong enough to boost a big workshop into space.

Forty-eight feet long and 21 1/2 feet in diameter, "the can," as it was affectionately called by the astronauts, was the largest inhabitable human-built object ever to be put into space. With the volume of a three-bedroom house, it weighed an Earth weight of almost 100 tons and when standing upright was as tall as a 12-story building.

Actually, in space, Skylab was a somewhat awkward looking cluster with five major assemblies fitted together. The largest piece was the orbital workshop, built from the unused Saturn 4B tank and turned into a double-level structure with separate living and working compartments. Making up the rest of the assemblage were an airlock module, fixed forward of the workshop (to make it possible for astronauts to move in and out of Skylab without disturbing the livability of the atmosphere inside), and a docking

adaptor which had provided a docking port for the Apollo command and service modules during Apollo missions. The cluster was completed by the Apollo Telescope Mount, which held a sophisticated solar observatory and additional power arrays, or "wings" of solar cells.

If not exactly the most streamlined and beautiful structure to be put into space, Skylab still had an important role to play. Besides turning toward the Sun to learn about its mysteries, it would also turn its instruments toward the Earth and from its high vantage point study Earth's resources, pollution problems, and weather patterns. Most importantly, though, it would act as a laboratory in which the astronauts themselves could be studied, a chance for science to take a close look at how well and how safely humans could work in space over extended time.

Launch date was May 14, 1973, and at first everything appeared to be going smoothly. The trusty

Saturn 5 did its job well and Skylab was projected neatly into orbit high above the surface of the Earth.

It was immediately obvious, though, that something was wrong. The "bird" was unhealthy. Vibrations from the launch had ripped off the meteoroid/thermal shield some 63 seconds into the flight, causing it, in turn, to tear away one of the two solar array wings and jam the other in a partially open position. With its protective shield gone and its power supply limited to the batteries in the Apollo Telescope Mount (ATM) and the partially deployed wing, Skylab was exposed to the Sun and heating up fast. Temperature on Skylab's outer surface climbed to over 300 degrees F.

NASA had a wounded bird on its hands, maybe even a dying one. As the internal temperature on Skylab climbed to 190 degrees F, it certainly looked as if the space station was doomed. Some sharp maneuvering by ground control was able to direct the craft out of its broadside position and oriented it in a more favorable angle to the Sun, cutting back on the ex-posure somewhat and positioning the ATM solar panels to collect enough power to keep the internal instruments operating. Whether the station could be kept operating or made safe enough to be inhabited by a crew of astronauts was a tough question.

Skylab SL-2: Astronauts to the Rescue

The first scheduled crew, astronauts Charles (Pete) Conrad, Joseph Kerwin and Paul Weitz, had been standing by for launch in their leftover Apollo com-mand/service module to join Skylab in orbit. With Skylab sick, though, the mission needed to be re-thought. A "fly-around" mission that would have had the crew scout the damage to Skylab and then return to Earth was quickly dismissed as a costly waste and Conrad, Kerwin and Weitz went to work with NASA and industry teams to try and come up with a solution to save the damaged craft.

Rehearsal for Rescue

Once launched, Skylab should have unfurled its solar "wings" to collect energy from the Sun and then quietly orbited Earth, everything ready and waiting for the first crew to pay a visit. Something, however, had gone wrong. One solar panel would not respond to commands to deploy.

Could a rescue mission find the problem and fix it while crew and Skylab orbited far above the Earth? Maybe—but the mission would require a lot of planning by NASA scientists and engineers. They had to deduce what was blocking the successful deployment of the solar wing and develop and test possible solutions. Every step needed to be well planned and rehearsed.

But how do you rehearse a complicated zero-gravity repair job while you are operating under the pull of Earth's gravity? The answer was to use NASA's neutral buoyancy tank, where, by adding weights, people and objects would float neither up nor down, simulating the weightlessness of space. This 1,300,000-gallon water tank at the Marshall Space Flight Center was 75 feet in diameter and 40 feet deep—large enough to hold full-size mockups of Skylab. Here, astronauts and engineers could float in and out and around the Skylab "look-alike" in much the same way they would float in the zero gravity of space. Prior to the launching of Skylab, the tank had been used by engineers and astronauts to test extravehicular activities, repairs and scientific experiments as they would be performed during missions.

Since there was no way to know at that point what was actually preventing the troublesome solar wing from deploying, the engineers tried a "pot-luck" approach and in the bottom of the tank they assembled a "junk pile" of possible blockages, including metal wire, bolts and other similar fragments of the failed meteoroid shield. Long-handled tools, like those used by telephone company lineman were fitted out with special flotation devices

15

attached to them. Floating freely in the tank to simulate neutral gravity effects, they were customized and adapted for any tasks at hand.

Astronauts Pete Conrad, Joe Kerwin and Paul Weitz then donned their pressurized space suits (outfitted with added lead weights to keep their buoyancy neutral) and entered the water tank. There—looking decidedly out of place at the bottom of the tank—Conrad, Kerwin and Weitz attacked the junk pile and practiced their space maneuvers as engineers and technicians watched and took notes. They pried loose, cut free and safely manipulated their way around the underwater section of the Skylab mock-up in what turned out to be a highly successful rehearsal for "the real thing" they would face a few days later during their critical mission in space.

At the Marshall Space Flight Center in Huntsville, Alabama, emergency teams tested out ideas and procedures in an underwater tank, used to simulate near zero-gravity and came up with a plan. The bird might be saved!

On May 25, 1973, Conrad, Kerwin and Weitz, now space-age emergency repairman as well as astronauts, finally were launched on board the Apollo command module for their Skylab rendezvous. On arrival at the stricken craft seven and a half hours later, a quick look confirmed Skylab's dire situation. Maneuvering around the damaged craft and televising pictures back to Earth, Conrad reported his observations to ground control and was given the go-head to attempt repairs.

The plan called for a parasol-type sunshade to be deployed by docking with the craft, climbing on board and pushing the parasol up and out through one of Skylab's airlocks. Once the "umbrella" was in place the shade would cool the craft down enough to be habitable so that other repairs could be made.

Before entering Skylab, though, it was decided that the astronauts would try and free the stuck solar panel. What had seemed difficult but possible in the underwater-tank tests on Earth, proved more difficult in space. Maneuvering the Apollo command module dangerously close to and around the stricken ship, Conrad operated the controls while Weitz stood in the open hatch with toggle cutters attached to a long pole and tried to cut free the jammed panel on Skylab. All three astronauts wore pressurized space suits designed to protect them during extravehicular activity (EVA) as Kerwin anchored Weitz down by holding on to his legs. Despite their efforts, though, the panel wouldn't budge. Abandoning the attempt after much frustration, Weitz climbed back inside the safety of the command module while Conrad prepared to dock their spacecraft with Skylab. There were more frustrations in store: The craft wouldn't dock. Finally, after six attempts, the problem was

discovered and 15 hours after launch a successful docking with Skylab was finally accomplished. With the command module and Skylab mated, the crew was at last able to take time out for some rest.

Operations went smoother the next day. When they climbed into Skylab the astronauts found that the heat had done less damage than they had feared. Though not comfortable, it was temporarily livable inside and they quickly deployed the parasol to bring the temperature down inside the ship. With Skylab at least in semi-working and living order the astronauts were able to begin a few of the planned experiments. Lack of adequate power was still a problem, though, and on the sixth day, after a full day of experiments, another power failure lost two more of the remaining 18 working batteries. Once again Skylab was in danger of dying altogether.

The crew would have to make another try at freeing the jammed solar panel.

On June 7, after elaborate planning and preparation, Conrad and Kerwin, swathed in their space suits, climbed outside and tried to free the blocked "wing." After another awkward and unsuccessful attempt using the pole and toggle cutters, Conrad slowly crawled along the outside of the workshop and lay his body precariously across the 1.2 m-wide and 9.4 m-long beam of the solar wing. From that hazardous position he was able to guide the cutters at close quarters while Kerwin, operating them by means of a lanyard from the workshop's "roof," severed the aluminum strap that was partially blocking the panel's deployment. Then, fastening a rope to the beam as they had carefully rehearsed, Conrad backed up and stood with the rope over his shoulder to gain leverage. With an abrupt lurch the beam snapped suddenly into place, almost sending Conrad and Kerwin flying off of the ship with a violent movement. The maneuver had been a success. Within minutes power was being fed from the freed panel into the

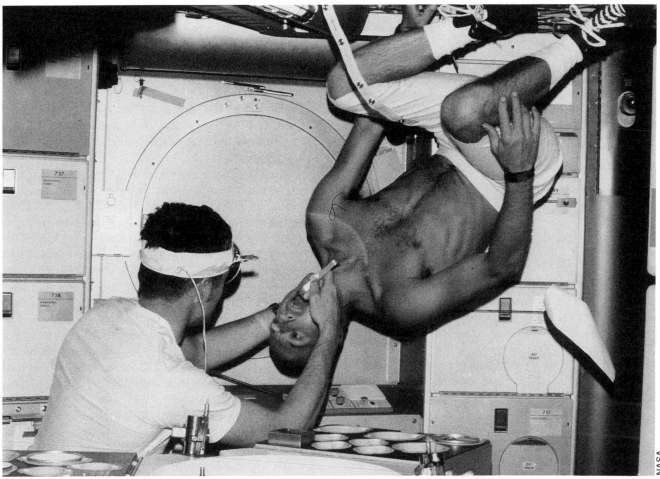

NASA

Scientist-astronaut Joe Kerwin examines Charles Conrad on board Skylab 2 mission

eight workshop batteries that had been useless since lift-off. Skylab was now completely "alive."

With everything up and running, the astronauts were able to devote full time to their work. Despite the early troubles and delays, by the end of their mission Conrad, Kerwin and Weitz had accomplished over 80 percent of their experiments and objectives. In a mission lasting 28 days, the first industrious crew of Skylab had managed to take 30,242 solar observatory photos and 8,886 Earth observation photos of 31 U.S. states and six foreign countries.

"We can fix anything!" Conrad had exclaimed confidently at the start of the journey, and in fact, they just about had. By giving a failed relay circuit to one of the failed ATM batteries a good whack with a mallet during a final one-and-one-half-hour space walk, at the end of their mission, Conrad even got one more holdout battery working. Returning to Earth on June 22, 1973, the first crew of Skylab had completed a highly successful mission demonstrating extraordinary ingenuity and skill.

The Second Crew Settles In

Launched July 28, 1973, the second Skylab mission (SL-3)—with Alan Bean, Owen Garriott and Jack Lousma aboard as crew—got off to a shaky start when two thrusters (engines) on the CSM proved faulty at the beginning of the mission. The problem proved to pose no serious danger for the crew, however, and rescue plans were cancelled as the three astronauts proceeded with what turned out to be a long mission filled with experiments and observation.

Bean, Garriott, and Lousma weren't alone though, as they arrived at Skylab. With them was a small zoo of traveling companions including six mice, 50 frogs' eggs, two minnows, 720 fruit fly pupae and history's first space-traveling spiders, Anita and Arabella.

One of the early jobs assigned to the second crew was to erect a new and more permanent Sun shield over the one deployed by Conrad, Kerwin and Weitz. The project had to be delayed, however, when the three crew members almost immediately began to

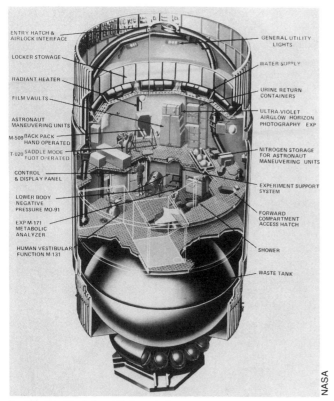

ENTRY HATCH &
AIRLOCK INTERFACE

LOCKER STOWAGE

RADIANT HEATER

FILM VAULTS

ASTRONAUT
MANEUVERING UNITS

M-509 BACK PACK
HAND OPERATED

T-020 SADDLE MODE
FOOT OPERATED

CONTROL
& DISPLAY PANEL

LOWER BODY
NEGATIVE
PRESSURE MO-91

EXP M-171
METABOLIC
ANALYZER

HUMAN VESTIBULAR
FUNCTION M-131

GENERAL UTILITY
LIGHTS

WATER SUPPLY

URINE RETURN
CONTAINERS

ULTRA-VIOLET
AIRGLOW HORIZON
PHOTOGRAPHY EXP

NITROGEN STORAGE
FOR ASTRONAUT
MANEUVERING UNITS

EXPERIMENT SUPPORT
SYSTEM

FORWARD
COMPARTMENT
ACCESS HATCH

SHOWER

WASTE TANK

NASA

Inside the Skylab orbital workshop

show signs of space sickness. Attacked by nausea and dizziness and unable to eat, Bean relayed their distress to Earth and the space walk necessary for deployment of the new shield was postponed for about a week.

The miniature menagerie wasn't faring much better. The mice and fly colonies had died after a control had short-circuited in their environmental capsules, and the disoriented minnows had begun to swim around in circles. Coping a little better than her co-passengers, Arabella started spinning her webs a bit erratically at first, but quickly oriented herself and began spinning in regular patterns, just as she would on Earth. (Could the bits of filet mignon fed her by the crew have helped?) When Anita also got a chance to spin, she adapted just as quickly.

After a few days' adjustment the crew began to feel better, although spells of space sickness would plague them off and on for weeks. Settling down to work, they began a demanding and rigorous schedule of photographing the Sun and Earth resources, running frequent health tests, checking equipment and performing other experiments.

The work went so smoothly for the crew that they often exceeded the daily work schedule. All three astronauts were particularly intrigued by the Earth

resources work and during one of their many sweeps across the face of the Earth they were able to communicate with a small fleet of fishing boats in the Gulf of Mexico and try out Skylab's instruments at spotting the most likely fishing areas based on water temperature and color. They also followed the life cycles of storms, looked for moist soil and water sources in Africa, watched the eruption of Mt. Etna, and made assessments of other soils, crops and mineral disbursements.

By August 6, the crew was feeling well enough to attempt the space walk and solar-shield erection. In keeping with the precedent that they had been setting with Skylab's scientific work, the space walk was a record breaker. In a six-and-one-half-hour EVA Garriott and Lousma managed to erect the new Sun shield by assembling two 55-foot rods and raising a new parasol over the one hastily deployed by their predecessors. Once this was done they went on to load the film canisters in the telescope cameras, survey the entire craft for damage and even check out the thrusters that had given them trouble on the Apollo command module. A solid day's work for a crew that only a few days before had wanted nothing more than to lie still for a while.

In fact, the crew's health had improved so greatly from the early days of their arrival, helped by a daily one-hour workout on a small stationary bicycle, that delighted NASA doctors Earthside gave permission for two more space walks near the end of the mission. On August 24, Garriott and Lousma, by now the "space-walkingist" of all the astronauts, moved out once more to plug in a new set of gyroscopes, replacing a finicky set of old ones, check Skylab over from stem to stern again and once more replace the film cassettes.

Then on September 22, Bean, who had been stuck inside as his buddies enjoyed their space romps, finally had his turn. Working alongside Garriott and getting his first taste of the view from outside the workshop, Bean, an Apollo Moon-walking veteran, was reluctant to return to the ship after the excitement and wonder of a stimulating two-and-one-half-hour EVA.

From their shaky start, Skylab's second crew had exceeded all the demands put upon them—and more. They spent over 300 hours on solar photography, 100 more than planned, and often working voluntarily 12 to 14 hours a day. They had even optimistically put in for an extension of their mission, but NASA officials, though delighted with their performance, decided it would be more prudent to bring them home as planned.

Closeup view of Skylab taken during "fly around" inspection by Skylab SL-3 crew

On September 25, 1974, Skylab's second hard-working crew splashed down in the Pacific, 230 miles southwest of San Diego, California. They had spent 59 days in space, a new record, including 13 hours and 44 minutes of EVA, had completed over 852 orbits and had traveled over 24 million miles (38.6 million kms). Returning with the triumphant crew were 75,000 pictures of solar activity, including six solar flares (explosive releases of energy on the Sun) and over 14,000 Earth resource photos. Once again, a Skylab crew could say to themselves—a job well done!

The Third Crew: Another Record-breaker

You could have made bets around Mission Control that no new records would be set on the third trip to Skylab. Launched on November 16, 1973, the three astronauts were the least experienced of the Skylab crews. Forty-one-year-old Marine Lieutenant Colonel Gerald Carr, civilian engineer Edward Gibson and Air Force Lieutenant William Pogue would find themselves being called "the grumpiest" of the three Skylab crew, but to everyone's surprise they would also become the most productive.

Things didn't start out that way, though. Still elated from their previous Skylab successes, NASA managers took a survey of supplies still on Skylab and planned an 84-day mission, the longest yet, with a

duty sheet for the crew that read like a Christmas "wish list" of observation and experimentation. The reasoning was, what had gone "great" before, could go "greater" now.

Not surprisingly, like the youngest child in a family of athletic or scholastic over-achievers, the crew of Skylab SL-4, the third and final mission, felt enormous pressure to succeed.

They reacted accordingly.

The griped. They grumbled. They dragged their feet. Pressured by an over-scheduled and over-optimistic workload the three astronauts often found themselves trying to do two or three things at the same time. Fatigued, not yet completely adjusted to the space environment and fighting off occasional bouts of nausea, they let off the steam by complaining.

The food was bland, the toilet too complicated, the exercise periods too long. They didn't like the brown color scheme and nothing worked right. What's more, they were getting tired of shaving for the television cameras.

But the planners and scientists on the ground kept adding to the workload. They seemed oblivious to the problems of the men who actually had to keep up with the grueling schedule of experiments, housekeeping, observations and equipment repair.

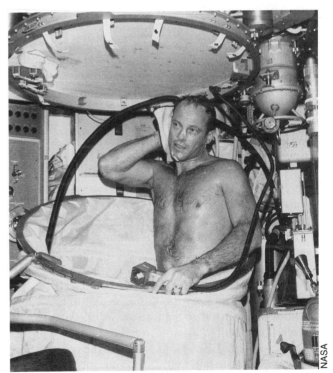

Skylab SL-3 crew member Jack Lousma takes a shower

When Sunday, usually scheduled as a day of rest, began to fill up with experiments and housekeeping, the astronauts reacted with what Carr later called "the first space mutiny." The three astronauts simply put down their tools and decided to the consternation of ground control to take the day off.

The situation both on Skylab and at Mission Control was becoming critical. As Edward Gibson later put it, "The first seven or eight days were not something that I would want to go through again."

Finally on December 30, 1973, things came to a head. In a precedent-breaking, frank and personal communication with the control center, Carr, Pogue and Gibson came out into the open with their gripes. They were men, not machines. They made mistakes and had to take time to correct them. There was too much activity being scheduled and not enough rest. They didn't even have time to look out the window. They were beginning to feel like slaves rather than partners, and it would be kind of nice to get some words of appreciation once in a while from the control team.

The frank confrontation cleared the air. Used to dealing with a more hard-headed approach to problems, Mission Control had suddenly gotten a more personal look at the human side of its space crew. Almost immediately the work schedule was readjusted and more leisure time allotted. Communications between control and the Skylab crew became more collaborative, and morale immediately began to improve all around.

Not surprisingly, as morale improved so did the mission.

Carr, Gibson and Pogue plunged into work with a new enthusiasm.

By splashdown on January 8, 1974, the third crew of Skylab had set a half-dozen new records. Their 84-day endurance record for the longest mission in space would stand unchallenged for the next four years. They also performed the longest single EVA in Earth orbit, lasting seven hours and one minute, and accumulated the most EVA time for one mission at 22 hours and 21 minutes. They set the new record for orbital distance traveled at 34.5 million miles and completed or surpassed all the experiments, tests and observations required of them, returning over 75,000 solar photos, and 17,000 Earth resource photos. During their mission they also studied and took over 2,500 photographs of the comet Kohoutek, experimented with crystal growth, and produced new metal alloys. They scored an additional coup when for the first time a Skylab crew was able to record a solar flare from beginning to end using the powerful instruments on board to study it in detail.

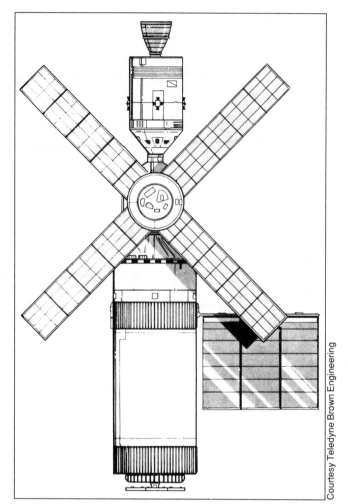

Skylab complex

Courtesy Teledyne Brown Engineering

As if to make up for lost time, Carr, Gibson and Pogue also managed to perform more exercises on Skylab's stationary bike and a portable treadmill, sending the results of more medical data back than either of the previous two crews. For a mission that had started so badly, Skylab 4 ended splendidly.

It was the last mission that Skylab would ever host.

The Death of Skylab

As the third crew left Skylab to return to Earth, Carr, Gibson and Pogue also concluded America's first era of manned spaceflight. Although the station's last visitors put together a welcome bag consisting of samples of food and drink, unused camera film, and clothing and electronic equipment for any future visitors, both astronauts and controllers knew that an era was over. Skylab had already begun to show signs of wear. No more immediate missions were planned. And funds—allocated by a Con-

The Salyut 7 space station before docking with Soyuz T-13

Skylab

The crew of Soyuz T-12 after landing

Artist's concept of the joining of Apollo and Soyuz (Apollo-Soyuz Test Project)

NASA

STS-1 launch: First Space Shuttle launch
(Columbia). April 12, 1981

NASA

Bruce McCandless approaches maximum distance from the Shuttle using the MMU, February
1984

Challenger accident January 28, 1986

Astronaut Sally K. Ride communicates with ground controllers from the mid-deck of the Earth-orbiting Space Shuttle Challenger

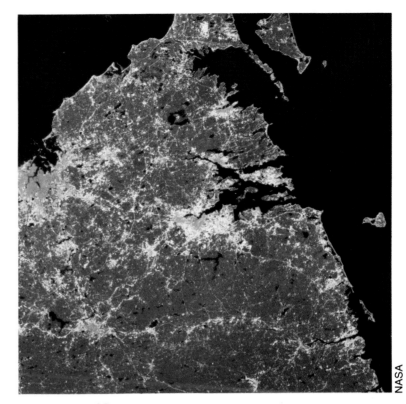

Landsat view of Boston

Landsat view of the border between Alberta, Canada and Montana in the U.S., showing the difference in land use

Proposed Mars Rover collecting samples on the Martian surface

Canadian astronaut shoulder patch

Computer simulation of aerodynamics at Mach 25 on the X-30, U.S. National Aerospace Plane (NASP)

NASA Ames Research Center

Communties of the future: orbiting space colonies

gress preoccupied with war in Vietnam and domestic unrest—were low.

To try and give the nation's first space laboratory some extra time before its orbit would decay, Mission Control used the thrusters on the attached command module to boost the station into a higher orbit just before the return home of the third crew.

The hope was that Skylab could stay in orbit for perhaps another nine or 10 years, long enough to be visited and put back into operation by the planned Space Shuttle.

Ironically, it was the Sun, the source of so much concentrated study by Skylab's crew, that hastened Skylab's early death. Increased sunspot activity in 1978 and 1979 caused the Earth's atmosphere to expand, creating increased drag, or pull, on the orbiting space station, combined with difficulties in holding the laboratory in a low-drag attitude (a position which could have minimized the atmosphere's effect), the problem cut short Skylab's planned lifetime. Delays in the delivery of the Space Shuttle system spelled the final death warrant for the lonely laboratory. After some sensational press and public concern about its predicted fiery reentry, Skylab plunged back to Earth on July 11, 1979, reentering over the Indian Ocean and scattering pieces over a thinly populated area of Western Australia. A lone outback jackrabbit was its only victim.

During its lifetime, Skylab completed over 34,981 orbits around our planet, over a period of six years and nearly two months. For a station that served so well and taught us so much about the Sun, the Earth and human abilities in space, Skylab sadly spent most of its time empty and abandoned. Of its nearly 75 months in orbit, less than six were given to human habitation.

How Do You Know Which Way Is Up?
Inside Skylab

On Earth, "up" and "down" are distinguishable. The floor is "down" by your feet and the ceiling is "up" over your head. But in space, because there is no gravity, feet may float in any direction and there really is no "up" or "down."

However, if an astronaut wants to be efficient when working at a computer terminal, using a camera or running an experiment, knowing which way is "up" becomes important. So workplaces (or workstations, as they're called) in Skylab were designed with foot restraints to hold the astronauts' feet in that station's "down" position while they worked.

However, engineers designing Skylab had tried to make the best use of the surface areas and saw no logical reason to coordinate up and down. In one compartment, there might be several "up/down" directions, depending upon how workstations and hatches fit most conveniently, and an astronaut moving through a hatch might see the ceiling of one compartment become the floor of the one he was entering. Once on board the astronauts found the up/down hodge-podge vaguely confusing. As Skylab SL-4 astronaut Gerry Carr put it, "There was always that little instant of disorientation. From a safety standpoint, if you're having a fire or an emergency of some kind, and you need to get to a critical control, you don't want to worry about disorientation as you're whistling through a hatch from one compartment to another."

Because Skylab was the first American space station, it became a laboratory for future spacecraft and space station design. Both Soviets and Americans have recognized since the days of Skylab that, ultimately, we are Earth creatures, and as such expect the sky to be light and the ground to be darker. Today, space station designs establish a distinct up and down with light paint on the "ceiling" and darker paint on the "floor." The system helps keep things from going totally topsy-turvy in space.

3

APOLLO-SOYUZ: ENDING AN ERA WITH A HANDSHAKE

Moscow is GO for docking; Houston is GO for docking. It's up to you guys. Have fun!
—Capcom Dick Truly
calling Apollo from Houston

Building the Foundation

It might have looked as if Hollywood was writing the script for NASA in 1975. Spurred by rivalry with the USSR, Americans had walked on the Moon and had also completed the last Skylab mission—the longest yet. Now, with those missions accomplished, for the United States the Apollo/Skylab era would end on a high dramatic note as Americans and Soviets met and joined hands in space—in the Apollo-Soyuz mission.

While some observers saw this historic link-up of the world's two great superpowers as a pure exercise in public relations, there were immediate practical benefits for both countries. Not the least of which was the fact that each nation would have a chance to get a closer look at the space hardware, management techniques and administrative capabilities of its rival.

The early groundwork for the historic union had actually been laid in the early 1960s between United States and Russian space scientists, a group that had always been a little more optimistic about such cooperation than their political counterparts. No definite projects were discussed, but these casual, if careful, exchanges eventually led to an agreement on space cooperation signed by Deputy Administrator of NASA Hugh Dryden and Anatoly Blagonravov of the Soviet Academy of Sciences in 1962. Although the agreement called only for the exchange of data from scientific and meteorological satellites, another agreement in 1965 called for a similar exchange of information about space medicine—the physiological effects of space on humans.

Agreements finally solidified into plans for a joint docking in space. After a complicated and ritualistic dance of meetings both in the United States and USSR, the agreement was formally signed by President Richard Nixon and Premier Alexei Kosygin on May 24, 1972.

With politics out of the way, or at least no longer center stage, both sides settled down to work out the technical difficulties of the flight. The early plans called for a joint docking between Apollo and the Soviet space station, Salyut, but this was almost immediately changed to an Apollo-Soyuz spacecraft-to-spacecraft link-up. Although various reasons were given for the changed plan, some Western observers believed that the Soviet schedule for adapting the Salyut space station to a two-port docking system had slipped, necessitating the change.

Whatever the reason for the change of plan, it soon became apparent that the Apollo-Soyuz docking (also known as the Apollo-Soyuz Test Project or ASTP) had

Artist's concept of the docking of Apollo and Soyuz

a number of technological bugs to work out. Since Apollo and Soyuz had different docking systems as well as different living environments for their crews, some modifications and compromises had to be made if the two craft were not only going to link up, but also allow an exchange of visitors.

The agreed-upon solution involved a special air-lock/docking module. Designed jointly by the United States and the USSR and produced in the United States by Rockwell International, the docking module (DM), 10 feet 4 inches long and 4 feet 8 inches in diameter, was an air-lock chamber and a two-in-one docking system, with one end compatible to hook up with the Apollo craft, and the other end compatible to Soyuz. The module, which also contained a communications inter-link as well as a small electric furnace for joint materials processing experiments, served as a passageway between the two craft. Since the Soyuz operated with an oxygen-nitrogen atmosphere at sea-level pressure and Apollo astronauts breathed a pure oxygen atmosphere at about one-third

that value, the DM would allow the crews to acclimatize themselves before moving between ships.

Besides the necessary hardware to be built or modified there were also crews and support staffs to be trained, communications and control centers to be prepared, countless other major and minor problems to be worked out and decisions to be made. That just about everything ran pretty much on schedule was considered a minor miracle, or at least an indication that neither nation cared to lose face by botching a deadline or assignment.

With over 44 exchange visits between working groups, held alternately in the United States and Russia in the four years before scheduled lift-off, a few minor problems were sure to arise, though. Whether some inconveniences were accidental or deliberate (visits to Russia by the astronauts were often scheduled by the Soviets for the cold Russian winter, and visits by the cosmonauts to Houston were usually scheduled by the Americans for the hottest summer months) was often a matter of debate on both sides.

Inside Apollo 18 and Soyuz 19

Both introduced in 1967, the Soyuz and Apollo spacecraft were each adapted for this special mission. Apollo had just two compartments on this mission: the command module, which provided crew quarters for the three ASTP astronauts, and the service module. A special 10-foot-long (3-m) docking module was also launched with the Apollo spacecraft and was maneuvered and locked into position on Apollo once in orbit. One end matched the Apollo hatch, while the other was designed to accommodate the Soyuz spacecraft.

As Apollo and Soyuz came together in orbit and docked (or "shook hands," as Leonov said), a two-day exchange of Soviet-American festivities in space commenced inside. The first day, which began with cosmonaut and astronaut reaching tentatively through the open hatchway, concluded with an American visit to the Soviet crew quarters in the orbital module. With five men in a compartment built for two, it was a tight squeeze as they huddled around the small green table and shared turkey and cranberry sauce, lamb soup and chicken. Leonov jokingly offered up tubes of borscht soup labeled as vodka. Posing for photographs, as well as sketches by Leonov, a talented artist, took up the remainder of that day's visit. Later in the mission, Leonov's portraits of the astronauts provided a stunning addition to the ASTP TV broadcast.

On the second day Slayton and Brand received a televised grand tour of the Soyuz craft, while Leonov crossed into Apollo with a casual "Howdy pardner," to join Stafford for a visit. Ever conscious of their vast TV audience watching below, both groups offered a few proud comments about their country's achievements. They punctuated the day's activities with a steak lunch aboard Apollo. The afternoon schedule offered a press conference during which Stafford and Leonov expressed enthusiasm for their successful cooperation. In Leonov's words, "We have found out that we can work together in space and cooperate... We have complete mutual understanding and everything we have planned we have accomplished."

Said Stafford: "How this new era will go depends on the determination, commitment and faith of the people of both our countries and the world."

There was also the question of security. Not surprisingly, though, as their respective countries worried over political problems, the astronauts and cosmonauts, who had learned each others' languages, appeared sincerely to enjoy each others' company. Shuttling between the two nations' training centers, the three American astronauts (Thomas Stafford, Donald "Deke" Slayton and Vance Brand) and the two Russians (Alexei Leonov and Valeri Kubasov) shared many of their off-duty hours together attending parties, hunting, water skiing and racing fast sports cars.

A Handshake in Space

On July 15, 1975, after four years of preparation, Apollo 18 and Soyuz 19 lifted off. Soyuz went first, at exactly 8:20 A.M. EDT, blasting off from the launch pad at Baikonur (Tyuratam) on the Central Asian steppes. Exactly on time, seven and one half-hours later, at 3:50 P.M. EDT at the Kennedy Space Center in Florida, Apollo began its pursuit. Since the United States spacecraft had a larger fuel tank and greater maneuverability it carried the docking module and was assigned the role of "active" partner. It would be up to Apollo to catch up, find Soyuz in orbit, maneuver into position and dock. To make actual physical sighting easier, Soyuz had been repainted a brighter color and fitted out with beacon lights.

Communications for the mission would be handled by a complex system that included 14 voice transmission lines, two television lines, two telex lines, two teletypewriter lines, and two datafax lines.

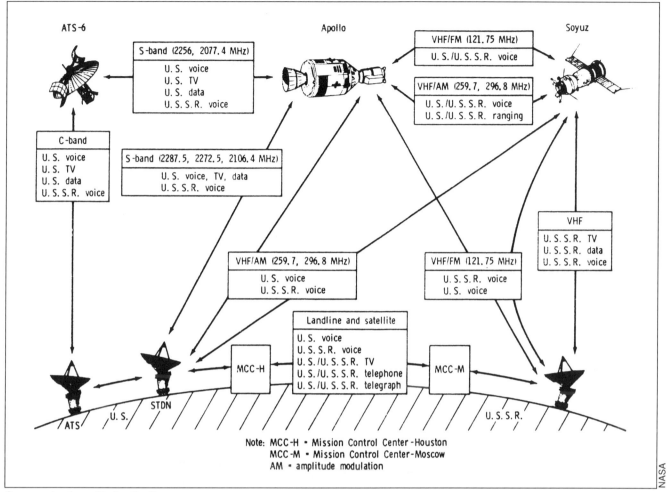

Communicating with Apollo-Soyuz

Communicating with Apollo-Soyuz

Coordinating communications for the Apollo-Soyuz mission was no small matter. NASA kept track of the Apollo-Soyuz mission using the Spaceflight Tracking and Data Network (STDN), consisting of both permanent and portable antennas on the ground, as well as specially equipped aircraft and ships carrying special instruments. The 14 STDN stations used by the United States were complemented by a Soviet system of seven ground sta-

tions and two ships. In addition, NASA was able to make use of its ATS-6 (Applications Technology Satellite) to relay U.S. and Soviet voice communications, as well as U.S. TV and data communications to the ATS ground station. Meanwhile, control centers in Houston and Moscow used both land lines and satellites to keep in touch with each other by radio voice communications, telephone, TV and telegraph.

To further help in communications between the two spacecraft and their control centers, NASA plugged its powerful communications satellite ATS-6 into the system. From its geostationary orbit, 22,300 miles (35,900 km) above the equator, the ATS-6 represented

the first time that NASA would use a spacecraft-to-satellite relay.

Forty-four hours and 35 minutes after the Apollo launch and 52 hours and 5 minutes after the Soyuz launch, the two spacecraft—lonely travelers in the blackness of space—found each other 100 miles above the Earth. At 12:09 P.M. EDT, July 17, under the control of Apollo commander Thomas Stafford, they docked.

"We have capture," announced Stafford from Apollo in Russian.

"Soyuz and Apollo are shaking hands now!" Cosmonaut Leonov added from Soyuz in English, picturesquely describing the two spacecraft docking.

The link-up of the hardware had gone perfectly, but the first human handshake between astronauts and cosmonauts was not so easy.

The problem started when Deke Slayton and Thomas Stafford began to open the hatch between Apollo and the docking module.

"Deke smells something pretty bad up in the docking module," Stafford reported back to mission control. "We're going to put on oxygen masks right now; and we're going to close that hatch!"

The astronauts reported that the smell was like burnt glue, immediately sending Houston into a check of their consoles! A fire on board could mean disaster for both crews, but Houston control could find nothing wrong. Fortunately the smell—probably from burnt velcro associated with an earlier experiment in the electric furnace—quickly faded. It had been a harried few moments though for the astronauts, whose minds must have raced back to the fatal tragedy of the Apollo 1 launchpad fire a few years before, when in a matter of just a few seconds three of their

fellow astronauts—trapped in the oxygen rich crew cabin—had burned to death.

Assured that there were no problems, Stafford and Slayton transferred to the docking module and increased its atmospheric pressure to about two-thirds sea-level pressure while the cosmonauts in Soyuz reduced their pressure to the same level. Without this slow procedure any attempt at moving between the two craft and their dissimilar atmospheric pressures would have resulted in problems much like the "bends" suffered by deep sea divers who come up to the surface too quickly from the ocean floor.

Finally, nearly three hours after docking, and after adjustments of equipment, television cables and cameras on both craft to give the watching world a better view, hatch number 3, between the docking module and Soyuz, was opened.

"Come in here and shake hands," Stafford laughed in English to Leonov waiting in Soyuz. "Alexei, our viewers are here," he added in Russian as Leonov stuck his head and shoulders into the docking module and into view of the television cameras. Then as the tension of the last hours broke, both men laughed and, as their linked spacecraft moved high over Amsterdam, they shook hands.

After four years of complex political and diplomatic exchanges, compromises, pessimistic warnings about "incautious technology transfers" and visionary dreams of "a new era in space cooperation," the two heroic space crews at last united.

During the next few days the crew of Apollo and Soyuz worked together, performed a few mutual experiments, visited each others' ships, communicated with each others' nations, received greetings and congratulations from politicians and statesmen, put on televised press conferences, had dinners together and performed another experimental docking.

A publicity stunt it might have been, but it was a spectacular one. For a few brief hours, five men from two nations who uneasily shared a small Earth lived together in unison at the doorway to the stars.

Near the Edge of Disaster: The Return Trip

The joint venture was over almost as soon as it had begun. On July 19, after a final exchange of gifts and farewells, the two capsules separated.

"Thank you very much for your big job," were the final words of Leonov as Soyuz gained distance from the watching Apollo and slowly drifted away. Returning home after a few more solo experiments, Soyuz carrying Leonov and Kubasov reentered the atmos-

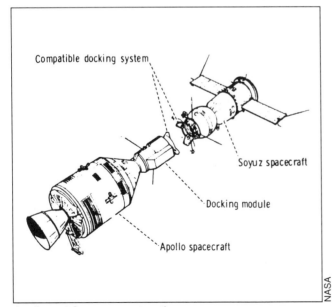

Compatible docking system

Soyuz spacecraft

Docking module

Apollo spacecraft

NASA

Apollo and Soyuz come together

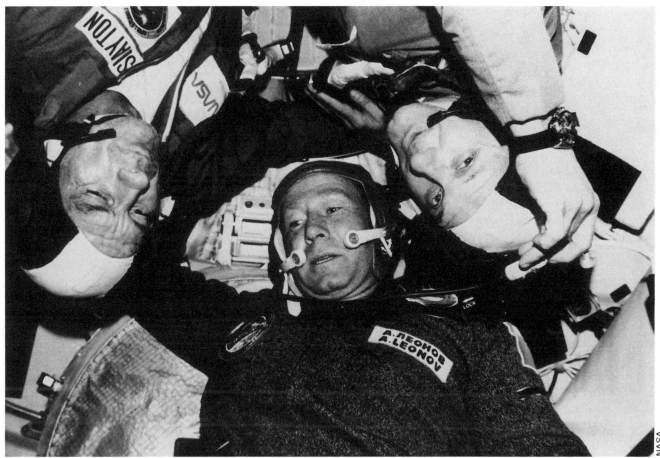

Deke Slayton (l), Alexei Leonov and Tom Stafford (r) get together aboard Apollo-Soyuz

phere and on July 21 parachuted to Earth near Arkalyk on the Russian steppes to be greeted by waiting crowds. Exhausted but elated, they had completed their mission. Now would come days of almost equally fatiguing ceremony and acclaim.

According to plan, Apollo spent a few more days in orbit, running down a last checklist of experiments and holding a final televised press conference. Like their Soviet colleagues, Stafford, Slayton and Brand knew that their mission had been a job well done. A routine reentry and splashdown for Apollo was scheduled for July 24, 1975.

Watching the televised landing in the Pacific 500 kms from Hawaii, millions of people who had witnessed the successful "handshake in orbit" relaxed as they saw a perfect splashdown. Unknown to them, at that very moment the three American astronauts were waging a battle with death.

During reentry, Brand had missed an instruction that Stafford was reading off their checklist, failing to turn on the Earth landing system (ELS) which at the proper altitude would deploy the parachutes. Slayton

noticed that the parachutes hadn't deployed at the right time and instructed Brand to deploy them manually—so that part of the problem was under control. However, the thrusters (engines programmed to fire to stabilize the craft) should also have been turned off by the ELS. Stafford immediately noticed this problem too, and cut off the fuel, but deadly nitrogen tetroxide oxidizer, an extremely poisonous gas, continued to be expelled and was drawn into the spacecraft through an open pressure release valve.

By the time Apollo hit the ocean "like a ton of bricks," in Stafford's words, the three men were coughing violently and struggling to retain consciousness. As Apollo turned upside down in the water, Stafford attempted to get hold of the oxygen masks, undid his safety straps and was thrown violently back into the docking tunnel. Crawling back out quickly he managed to get hold of the masks—but by then Brand was already unconscious. Slayton and Stafford donned their masks and struggled to get a mask on Brand. In less than a minute Brand came around, but the crew was still sealed inside the bobbing capsule.

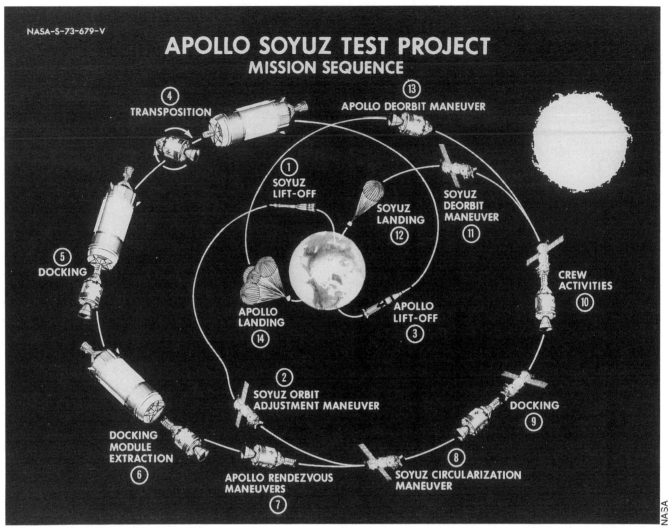

NASA-S-73-679-V

APOLLO SOYUZ TEST PROJECT
MISSION SEQUENCE

④ TRANSPOSITION

⑬ APOLLO DEORBIT MANEUVER

① SOYUZ LIFT-OFF

SOYUZ LANDING ⑫

SOYUZ DEORBIT MANEUVER ⑪

⑤ DOCKING

CREW ACTIVITIES ⑩

APOLLO LANDING ⑭

APOLLO LIFT-OFF ③

② SOYUZ ORBIT ADJUSTMENT MANEUVER

DOCKING ⑨

DOCKING MODULE EXTRACTION ⑥

APOLLO RENDEZVOUS MANEUVERS ⑦

SOYUZ CIRCULARIZATION MANEUVER ⑧

NASA

Apollo-Soyuz Test Project mission sequence

In the general uproar outside, as frogmen, ships and helicopters converged on Apollo, the shouts and frantic banging of the three astronauts went unnoticed until an angry Stafford sprung the latch himself and the three managed to struggle to fresh air.

After heroically holding an animated 10-minute conversation with President Gerald Ford from aboard the recovery ship, the three men reported the accident at the routine post-recovery medical examination and were quickly shipped to Honolulu for further tests. All three had suffered some ill effects and Slayton's lungs had been blistered, but after a few nervous days, during which wives and children were flown to visit the confined astronauts, the prognosis was good. No permanent damage was expected. Ironically Deke Slayton, who as one of the original Mercury 7 had once before been grounded for medical reasons, was found to have a small shadow on his left lung that had actually been overlooked before the flight. Fortunately it turned out to be a benign lesion and not, as had been feared, cancerous.

A Look Back at Apollo-Soyuz

Had Apollo-Soyuz been worth all the hoopla?

Certainly, to the gigantic crowds that turned out to welcome the joint visits of cosmonauts and astronauts in both countries in the following weeks, it seemed so. The public relations aspect of the mission had been a tremendous success. In addition to the engineering data returned by the mission, some science was done by both teams—they had made observations of Earth, examining deserts, icebergs and oceans. The Apollo crew examined the near-Earth atmosphere and tracked the jettisoned docking module to measure

Earth's gravitational field. In addition, 11 experiments aboard Apollo had added 125 hours to materials processing experiments (for example, growth of crystals and formation of alloys in zero-gravity) accumulated aboard Skylab.

ASTP's Apollo 18 would be the last to fly in a program that saw its spacecraft carry 45 astronauts in seven years, including 15 flights into Earth orbit, and six landings out of nine visits to the Moon. Apollo-Soyuz represented the end of an era, the last of manned United States spaceflights before the Space Shuttle was launched—and a new era would begin.

4

SALYUT 6 AND 7: THE CABIN BECOMES HOME

*S*pace is very beautiful. There was the dark
velvet of the sky, the blue halo of the Earth and
fast-running lakes, rivers, fields and cloud
clusters. It was dead silence all around...The
panorama was very serene and majestic.
                                    —Valentin Lebedev

The late 1970s and '80s saw a consistent broadening of the Soviet experience in space as they steadily upgraded the design of their outposts, improved their computer systems and revised the living conditions, physical regimen and lifestyle of their cosmonauts. The result would be a prodigious gain in the sum total of their experience in facing the rigors and solving the problems of living and working in space.

Success Begins to Beckon: Salyut 6

Salyut 6—a new design with many added improvements—made its debut on September 29, 1977. Perhaps the biggest breakthrough for the new station was the addition of a second docking port. The second port created two big advantages: It permitted a Progress cargo ferry to dock while a Soyuz spacecraft remained at the other port, and it made visiting possible by a second Soyuz crew.

Basically the new Salyut looked much like its predecessors, but it featured a major innovation—the Delta computer system. This new automatic orientation and control system relieved the crew of routine control tasks and freed up large blocks of time for more scientific experimentation. A new propulsion

system made refueling in orbit possible. So now, when the drag of Earth's atmosphere began to pull Salyut down from its orbit (which uncorrected would cause the same kind of orbital decay that brought down earlier Salyuts and the U.S. Skylab), the Soviets had the extra fuel to fire Salyut 6's engines to boost it to a higher orbit. Introduction of a new water-regeneration system made it possible to reuse water captured from the station's internal atmosphere, while designers addressed the problem of cosmonaut comfort with a new shower stall that sprayed hot water under pressure into a waterproof plastic cabin. (The shower, unfortunately, proved unpopular with the crews because of the hours required to mop up stray water droplets and fold up the system). In addition, Salyut 6 sported three interior TV cameras (two black-and-white and one color), three black-and-white TV cameras on the station's exterior, a teleprinter and a new space suit for space walks.

Salyut 6 was designed to stay in orbit for a relatively long time; in fact, it stayed aloft for four years, 10 months, not giving up until July 29, 1982, and greatly outlasting the lifetimes of its predecessors. Of the 20 Soyuz missions that would attempt to dock with

Salyut 6 over the coming four years, 18 would succeed (carrying a total of 33 cosmonauts—three of them twice—for a total of 676 days in space). Eight of its visitors would be cosmonaut researchers from non-Soviet countries. Salyut 6 was the first station to approach the Soviet goal of a permanent presence in space.

With Soyuz 25, however, the new era was off to an unpromising start. Launched 10 days after Salyut 6, what was to be an exciting initiation of the new station ended in yet another disappointing docking failure.

Nine weeks later Yuri Romanenko and Georgy Grechko made good aboard Soyuz 26. In what is sometimes called "The Classical Mission" (because their names mean "Roman" and "Greek" in Ukrainian), they packed their stay full of adventures— remaining aloft for 96 days, 10 hours and at last breaking the U.S. Skylab world record of 84 days in space (November 1973 to February 1974).

In the first Soviet extravehicular activity since 1971, Grechko checked out the docking port that had apparently caused Soyuz 25 problems. (Soyuz 26 had docked to the second port.) Finding everything in order, Grechko was taking a quick look around to enjoy the awesome expanse of space when he unexpectedly caught sight of his partner Romanenko floating by— his safety tether accidentally hanging loose! Grechko's quick lunge to catch the passing figure narrowly averted tragedy and a painful death for Romanenko.

The danger wasn't over, however. Once they'd floated back inside the airlock, the two cosmonauts closed the hatch and prepared to repressurize that area so they could open the opposite hatch leading to the station and return to the safety of its life-support systems. But a reading told them that the depressurization valve was stuck open! Their EVA supply of oxygen was dwindling fast. Reviewing their few options with ground control, they finally decided to try to repressurize the transfer compartment anyway in the hope that the reading was a false alarm. They literally breathed a lot easier to find that this was the case. All went well with the repressurization and they climbed back into the station.

Progress—The Cargo Ferry

First Launch: January 20, 1978 (to Salyut 6)
Total Weight: 15,476 lbs. (7,020 kg) including payload
Payload Capacity: 5,071 lbs. (2,300 kg) or 30% of lift-off weight—2,025 lbs. (1,000 kg) fuel and 2,866 lbs. (1,300 kg.) dry cargo
Dimensions: 7.2 ft. (2.2 m) x 26.2 ft. (8 m)
Flight Time: Eight days

The Progress is an automated Soviet cargo ferry designed along the same lines as the Soyuz. Since its first launch in 1978 it has become a key player in Soviet long-duration missions, providing crews aboard Salyut and, more recently, Mir with much-needed supplies of food, compressed air and nitrogen, fuel, scientific instruments and other eagerly awaited necessities. Even the mail arrives aboard Progress, which docks at the second port, with the on-board crew's Soyuz craft docked at the first.

Since it has no solar panels, Progress is limited by its batteries to an eight-day flight, but it can remain hooked up to the space station for several months. Its empty cargo holds a convenient catch-all garbage can. Before sending the multi-use craft on its way back to the Earth's atmosphere, where it incinerates on reentry, the crew can boost their orbit using the Progress thrusters.

Progress is slightly longer than the Soyuz because it carries an extra instrument module, but launches easily since it carries no launch escape module, no crew and no heavy heat shield.

Carrying three external lights and two external TV cameras, the Progress sends rendezvous data both to ground mission control and to the space station crew, who can help out with automatic docking if needed.

The Social Graces in Space

Being enclosed with another person in a tiny area 10 feet by 25 feet, with no opportunity to take a night off at the movies or even just step out of the back door for a breath of fresh air, can get on the nerves of even the most easy-going personality. By the time Salyut 6 was designed, Soviet psychologists had put a lot of effort into making the environment as pleasant as possible. Soft pastel colors made the quarters more homey. A painted-on "ceiling" of lighter color contrasted with the darker "floor" to create the illusion of up and down, and the Soviets toned down some of the on-board clatter that caused cosmonaut Georgy Grechko to complain following his Salyut 4 stay in 1975.

Some of the major adjustments were social—and had to be made between the members of each team to avert the nightmare of short tempers and ill feeling. As Grechko explained in an interview following his Soyuz 26 flight with Yuri Romanenko, the two had made some key decisions from the outset. First, to eliminate possible resentment, there would be no leader—no one ordering the other. Second, they tackled the problem of unpleasant tasks. The official system called for taking turns, but the two decided to do them together. As Grechko explained: "Neither would wait for the other to start a routine job, say, changing the toilet filters—but, rather, he would try to do it himself." Seeing the work begun, the other would join in to help. Similarly they shared responsibility for mistakes and mishaps.

In such close quarters, probably the greatest problem was lack of privacy. Great care was required not to step on the other's psychological territory, as Grechko explained: "When I was going to say something, I first looked to see what mood he was in. Perhaps he was busy or deep in thought—it meant I should wait. Another time perhaps I might remember something funny at a time when he was gloomy—I felt I should keep my joke until the proper moment."

Putting the second docking port to work, Grechko and Romanenko received the first shipment of provisions and fuel aboard the unmanned Progress 1 ferry, launched January 20, 1978, and they received historic week-long visits from two crews of cosmonauts, traveling aboard Soyuz 27 (January 11-17) and Soyuz 28 (March 3-10). In yet another Salyut first, Grechko and Romanenko swapped spacecraft with the Soyuz 27 crew, who took the Soyuz 26 home instead of their own craft, showing that a second Soyuz could act as a relief mission or provide emergency transportation home for the main crew. Soyuz 28 arrived with the first non-Soviet, non-American space traveler, 29-year-old Captain Vladimir Remek of Czechoslovakia.

Beginning March 12 Grechko and Romanenko began a special pre-descent regime involving the use of a newly designed Chibis "vacuum" or low-pressure suit to force the cardiovascular system, grown lazy by weightlessness, to work harder by creating negative pressure on the lower part of the body. The "atmosphere" pumped into the trousers of this suit was actually of much lower pressure than we're used to from the atmosphere on Earth. As a result, the cosmonauts' blood flowed more readily into the legs from their upper bodies. On Earth, gravity pulls blood into our legs, and our hearts must work to pump it back to the upper body. The Chibis suit gave the heart a similar workout through use of reduced ("negative") air pressure. In addition, cosmonauts increased workouts on the running track and added exercises for the lower limbs and the blood vessels.

March 16, 1978, Georgy Grechko and Yuri Romanenko climbed into the Soyuz reentry module, closed the hatch, and headed home, touching down without a hitch. They arrived heroes, with Grechko holding the world record for total time in space at 126 days, between his Salyut 4 mission (aboard Soyuz 17) and his Salyut 6 stay.

Three months later, on June 15, 1978, the second extended-stay crew headed for Salyut 6 in their Soyuz 29 vehicle. During their mission Vladimir Kovalenok and Alexandr Ivanchenkov, both in their mid-30s, entertained two Progress supply wagons and two more crews of visitors (Soyuz 30, commanded by Pyotr Klimuk, and 31, commanded by Valery Bykovsky), including two non-Soviet cosmonauts—a Pole, Miroslaw Hermaszewski, and an East German, Sigmund Jaehn. The Soyuz 29 crew's 139-day stay set yet another record, and they returned home November 2 on the swapped Soyuz 31.

The third long-duration team, launched February 25, 1979, aboard Soyuz 32, continued the Soviet assault on the problems of human endurance in space. Vladimir Lyakhov and Valery Ryumin set an impressive new endurance record of nearly six months in space, living in an area scarcely larger than a 10-foot-by-25-foot living room.

To make the stay all the lonelier, the two projected visits from other cosmonauts fell through. A major engine failure on Soyuz 33, just as it was about to dock with Salyut, forced an immediate turnaround and descent. The following investigation postponed the next planned foray, leaving the two record-setters alone for their full tour. During that time, they conducted observations using the K-T-10 33-foot (10 m) radiotelescope and worked in an unscheduled extravehicular expedition to cut loose the telescope, which had become snagged when they had tried to stow it. The two returned to Earth aboard Soyuz 34, which had been launched unmanned. No manned mission would return to Salyut 6 for a full year.

Soyuz T Makes Its Debut

Meanwhile, however, Soviet engineers had been hard at work on a new Soyuz design—the T-series spacecraft. Much like the previous craft, the key difference was that now Soyuz could once again carry a troika or trio of cosmonauts. Two early, unmanned test models, designated as Cosmos 1001 and Cosmos 1074, were launched into orbit, followed by Soyuz T-1, also launched unmanned on December 16, 1979. It docked on the unmanned Salyut 6 the next day and stayed for 101 days, then detached by remote control. After the arrival of Progress 8 with fuel, spares and supplies four days later, Salyut was ready once again for company.

Strangely enough, and quite by accident, the fourth extended-stay mission to Salyut 6 also included Valery Ryumin—putting him in space almost a full year's time

between February 1979 and October 1980. Because the scheduled cosmonaut had broken his knee, Ryumin became the hands-down space endurance record-holder (a distinction he held until 1986) with his grand total of 361 days, 21 hours and 34 minutes logged on three different flights.

Arriving at Salyut aboard the conventional Soyuz 35, Leonid Popov and Ryumin would receive visits from four Progress cargo ferries and four teams of cosmonauts, including Hungarian cosmonaut Bertalan Farkas with Valeri Kubasov aboard Soyuz 36, the first team of short-term visitors to Salyut in two years.

Three days after the departure of Farkas and Kubasov, the first manned Soyuz of the new T-series arrived on June 6 with just two cosmonauts, Yuri Malyshev and Vladimir Aksenov, aboard. Two more visits included Vietnamese cosmonaut Pham Tuan, with Viktor Gorbatko, arriving on Soyuz 37 and returning on Soyuz 36, and Cuban cosmonaut Arnaldo Mendez and Yuri Romanenko aboard Soyuz 38. Popov and Ryumin ended their stay on October 11, returning on the fresh Soyuz 37 craft left behind by their visitors.

In a brief 12-day test flight Leonid Kizim, Oleg Makarov and Gennady Strekalov tried out the three-man capacity of the new Soyuz in the Soyuz T-3 trip to Salyut on November 27, 1980. After taking care of a few maintenance tasks—installing a new temperature control pump and a new fuel system converter—they returned to Earth.

The last long-stay crew to visit Salyut 6 arrived at the three-and-a-half-year-old station aboard Soyuz T-4 on March 13, 1981. With film and instruments taking the place of a third cosmonaut, Vladimir Kovalenok, commander, back for his second stay aboard Salyut, and Viktor Savinykh, a specialist in aerial photography and mapping, made up the crew. During their 74-day stay, the pair took photographs, performed experiments transmitting three-dimensional images through the uses of lasers (holography), and tended three greenhouses in a continued effort to grow plant life in microgravity.

They were joined on March 23 by Vladimir Dzhanibekov and Jugderdemidyn Gurragcha of Mongolia aboard Soyuz 39, and during the seven-day visit the two crews performed extensive materials processing and observed and photographed Mongolian territory using Mongolian equipment.

After a full schedule of 30 experiments during their stay, the Soyuz 39 headed home with containers full of processed materials, hologram film, tadpoles hatched in space and Salyut 6's greenhouse produce.

Soyuz 36, a joint Soviet-Hungarian spaceflight, getting ready to launch

Eastfoto

On May 15 Kovalenok and Savinykh received their last set of visitors, Leonid Popov, on a second visit, and Romanian cosmonaut Dumitru Prunariu. Arriving aboard Soyuz 40 for a seven-day stay Popov and Prunariu took the original Soyuz design out for its last spin. While on board, the crews performed studies of the upper atmosphere prepared by Romanian scientists, then wrapped things up, with Popov and Prunariu returning aboard Soyuz 40 on May 22, 1981, followed six days later by Kovalenok and Savinykh on Soyuz T-4. Twenty years after Yuri Gagarin's first spaceflight in 1961, Savinykh was the 100th human being and the 50th Soviet cosmonaut to fly in space, as the era of Salyut 6 came to an end.

During its three years, five months, of active operation, Salyut 6 had seen five long-duration missions, with the cosmonauts performing a total of 1,600 scientific experiments, more than 800 of them biomedical.

They had taken a hefty 15,000 photos covering more than 48 million square km of the Earth's surface. And with their 676 days of occupation they had far surpassed the U.S. Skylab record total of 171 days in 1973-74. They had also broken three successive individual space-duration records.

The grand finale for Salyut 6 took place when a large spacecraft called Cosmos 1267, which had been launched on April 25, 1981, docked with the unmanned space station on June 19. The resulting 31-ton (34-metric-ton) spacecraft was then tested extensively for safety and maneuverability—probably as a forerunner for future, larger space stations to come.

The Seventh Salyut

With Salyut 6 phasing out, Salyut 7 appeared on the scene with an April 19, 1982, launch. Sporting the same basic design as Salyut 6, the new space station did, however, include several improvements. Structurally, larger and stronger docking ports meant that larger craft could come visiting. The solar panels and batteries provided more power, and new protective screens had been added to the portholes to prevent damage from the deluge of micrometeorites that showers through space. Meanwhile, new pressurized suits for extravehicular activity (EVA) meant cosmonauts could now spend up to five and a half hours outside the station.

On board, an improved computer system left the cosmonauts more time for scientific work instead of spending hours at a time calculating maneuvers and orbits for the station. Salyut 7 was also fitted out with several new pieces of scientific equipment. For biomedicine, the new "Aelita" biomedical device hooked up to the inboard computer for improved spaceborne studies of the human vascular system and brain. Industrial research would profit from new furnaces for manufacturing semiconductors in microgravity and for experiments on the fusion of dissimilar materials. For astrophysical research an X-ray telescope (which detects X-rays from space) was added to the gamma-ray telescope (for study of extreme short-wavelength radiation from space) previously used on Salyut 6. Both types of telescopes are only useful when carried high above the Earth's atmosphere by a spacecraft like Salyut or a satellite.

Planners had also made several efforts to make the Salyut more homey for the crews. Redecorated with bright colors, the living quarters boasted a new refrigerator and a Rodnik ("spring") hot-and-cold water system. Partly in response to complaints by Grechko about the noise on board, much of the machinery noise was either moved away from the living area or was muffled to reduce annoyance to the crew.

Unlike the previous meal system, menus for the Salyut 7 cosmonauts were not planned before launch. Instead the crews could choose from among several possibilities, and could just request new supplies to be sent up on a Progress cargo shipment when they ran low.

Typically, breakfast might include canned ham, white bread, cottage cheese with jam, cake, coffee and vitamin pills. Dinner (or lunch) might consist of vegetable soup with cheese and biscuits, canned chicken, plums with nuts and the usual vitamin pills. The cosmonauts then might look forward to a supper of canned steak, black bread, cocoa with milk and fruit juice. Frequently, visiting crews or Progress shipments would bring along fresh vegetables, bread, and other delicacies from home that would considerably perk up the regime.

The first cosmonauts did not arrive until just under a month after the Salyut launch, on May 14 aboard Soyuz T-5. Its two-man crew, Anatoly Berezovoy and Valentin Lebedev, would set the first of three long-duration records aboard Salyut 7, with a stay of 211 days. They would be visited by two Soyuz crews—aboard Soyuz T-6 and T-7—and three Progress ferries. During their flight they staved off longings for home with tape recordings of the sounds of Earth—rain falling, birds singing, the rustling leaves on a summer morning, and families sent "home movies" to amuse them. However, homesickness and irritability did prove to be a problem for the cosmonauts. As Lebedev later said, though, "...no eruption must be allowed, for a crack, if it appears, will grow wider." In an August 1983 *Pravda* interview, Lebedev also described how he tried to diminish his boredom and anguished longings for home and family by gazing at Earth through a porthole.

Even if the psychological problems of living in space were still troublesome, however, Soviet medical scientists declared that the physical reactions of Lebedev and Berezovoy showed that artificial gravity would very likely be unnecessary aboard interplanetary ships of the future. The human body, they asserted, could withstand long periods of weightlessness if precautions were taken.

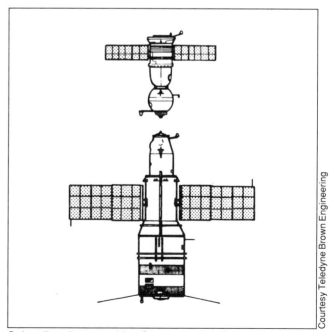

Salyut 7 and approaching Soyuz spacecraft

Courtesy Teledyne Brown Engineering

Research projects begun on Berezovoy and Lebedev during their "initiation" stay on Salyut 7 spanned a broad range of science, including biology, forestry, agriculture, geology and oceanology. Shortly after arrival, on May 17, they launched a small amateur radio satellite, the Iskra 2, designed by students in Moscow.

On June 25, 1982, about a month after resupply by a Progress cargo ferry, the crew of Soyuz T-6 was knocking at the Salyut 7 door. Vladimir Dzhanibekov, on his third space station visit, and Alexander Ivanchenkov, on his second, had brought with them Jean-Loup Chrétien of France, the first Westerner to travel aboard a Soviet spacecraft. As they approached Salyut a major computer failure onboard Soyuz T-6 made manual docking necessary. On return home Chrétien described the crisis, explaining that "following the computer breakdown the vessel began to rotate about all three axes, and was thus like a stone rolling over. We had to act very quickly." However, Dzhanibekov took over and masterfully brought the wayward Soyuz into port, and their arrival signaled yet another first, with five people aboard a Salyut station at one time.

The visitors stayed for a week, working at a grueling pace on numerous complex experiments in space medicine, materials processing, biology and astronomy. The schedule was so demanding, in fact, that in order to complete it they had to cut back on sleep. Seventeen French laboratories had contributed projects in a variety of areas, including a device called the Echograph (sent up on the Progress), which used ultrasound probes to study the functions of the heart and the circulatory system without the force of gravity working against it. The crew also did materials processing in the furnace and studied galactic objects and dust clouds in interplanetary space and the upper atmosphere.

With the Soyuz T-6 craft stocked with data tapes, film and processed materials, the crew began their trip home. Chrétien would later describe the 21-feet-per-second (6.5 m/sec) descent, slowed to 11.5 feet per second (3.5 m/sec) by retrorockets (which fired against the direction of flight) just before landing, as much more dramatic even than the launch. Everything progressed to a smooth touch-down on July 2, and French scientists seemed pleased with the results despite strained political relationships between the two countries.

Back aboard Salyut 7, Berezovoy and Lebedev, having given their colleagues a send-off, were exhausted. They enjoyed a brief rest period, after which Lebedev did a space walk to replace a device for measuring micrometeorite strikes and to set out some sample panels of structural materials to test the effects of their exposure to the rigors of space.

After a resupply visit from Progress 14 they were all set for their next visitors, who this time would include the second woman in space. Nineteen years after the historic flight of Valentina Tereshkova, her first successor, Svetlana Savitskaya arrived with her colleagues Alexander Serebrov and Leonid Popov aboard Soyuz T-7 on August 19, 1982, for an eight-day mission in space. The 34-year-old former sky diver and world aerobatic champion was known for her spectacular flying feats and came to the space program with 1,500 hours of flight time to her credit as pilot of 20 different types of aircraft. She may well have been hurried into the cosmonaut lineup by the Soviets to get the best of United States plans to send astronaut Sally Ride (who would be the first American woman in space) on a Shuttle flight in 1983. But in any case the moment was historic. Not only did Savitskaya break the long boycott against female space travelers, but she also demonstrated that women are as well (or as poorly) suited physically for spaceflight as men.

On her arrival, the resident crew greeted Savitskaya with flowers and an apron, joking whimsically (and perhaps hopefully) that she had drawn the cooking detail. Her reply, however, came quickly. "Housekeeping details," she asserted firmly, "are the responsibility of the host cosmonauts." On their return trip Serebrov, Popov and Savitskaya used the Soyuz T-5 spacecraft, leaving the fresh Soyuz for their hosts.

After the departure of their visitors, Berezovoy and Lebedev settled in for a long session of materials processing as they passed the 185-day space-duration record mark on November 14. They used a sort of spaceborne mini-production plant called Korund, which fed capsules through a revolving drum system into an electric furnace. With it the cosmonauts were able to produce highly regular crystals unlike anything that could be produced under the pull of gravity.

The two returned to Earth aboard Soyuz T-7 in a difficult night landing plagued by fog, low clouds and snow, which forced their rescue helicopter into a crash landing before it reached them. A follow-up helicopter and cross-country vehicles finally succeeded in rescuing the two exhausted cosmonauts, miraculously enough, only 40 minutes after their landing.

Meanwhile, back at Salyut 7, the furnace continued processing in zero gravity—without the vibrations and disturbances caused by the presence of humans.

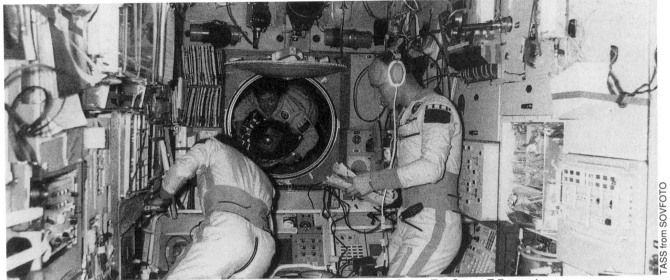

TASS from SOVFOTO

Berezovoy, Lebedev and Savitskaya (in transfer hatch) aboard the Salyut 7-Soyuz T-5-Soyuz T-7 complex

Getting Things Back from Space

One problem still not solved by the Soviets, however, was how to get processed materials back from space in bulk—a problem to reckon with if humans wanted to make manufacturing in space truly viable.

Then, just three months after the departure of Berezovoy and Lebedev, the Soviets sent a new visitor up to Salyut 7. Cosmos 1443, an 18-ton (20 metric-ton) space station-module/space tug, was launched March 2, 1983, and, with a slow, eight-day approach, docked on March 10. On board, it carried three tons of materials for use by the next Salyut 7 crew. At about 42.7 feet by 13 feet (13 m by 4 m) this tug was not just a bigger Progress. It could also make a return trip—using a reentry module that could carry a 500-kg package back to Earth for recovery at sea. The Soviets, it seemed, had found a method for bulk return transport of materials such as zero-gravity-processed materials developed aboard Salyut, as well as exposed film, used hardware and parts for repair.

In addition, Cosmos 1443 carried a large load of fuel that could be transferred back and forth between its tanks and Salyut's, making for increased maneuverability. Also, unlike the Progress, the new tug carried solar panels for additional power.

On April 21 Soyuz T-8 attempted to dock to the Salyut 7/Cosmos 1443 complex, but failed because of problems with the radar system and had to return to Earth. However, Soyuz T-9 succeeded on June 27, with Vladimir Lyakhov and Alexander Alexandrov aboard. The three-vehicle complex—with Salyut 7 sandwiched between Soyuz T-9 and C1443—had a total weight of 42.6 tons (47 metric tons) and a length of 115 feet (35 m). Unfortunately, however, the complex proved unwieldy to maneuver, with only semi-automated navigation, and problems developed. The Cosmos pulled away from Salyut 7 on August 14, divided in two to send its reentry module back to Earth for retrieval on August 23, and finally reentered the atmosphere itself on September 19.

Salyut 7 Develops Problems

In July, early in their stay, Lyakhov and Alexandrov experienced what they called an "unpleasant surprise" when a loud crack resounded over the usual hum of equipment. Something—manmade debris or micrometeorite—had struck a porthole, leaving a .15-inch (3.8-mm) crater but no real damage except to the crew's peace of mind.

That peace of mind was in for further distress in early September when a main oxidizer line broke, spewing two-thirds of the station's fuel out into empty space, damaging the control system and threatening the safety of Salyut. The crew retreated to their space suits and the safety of Soyuz T-9 until they received an OK from ground control to go back into the station. Once the danger was past they returned calmly to their work, however, concentrating on processing semiconductor materials and producing biologically pure substances for medicine, including molecular flu vaccines. After all, help would be along soon.

But help, in the form of what should have been Soyuz T-10, suddenly exploded into flames. For cosmonauts Vladimir Titov and Gennady Strekalov, who had also been aboard the unsuccessful T-8 mission, what started out in part as a rescue mission for the T-9 crew turned into near-disaster for themselves. As their launcher neared lift-off time on September 26, 1983, fire suddenly broke out at its base, rapidly spreading to the wiring, where it prevented the automatic abort signal from reaching the escape tower through the electrical system. Launch Control came to the rescue by sending the signal by radio, however, putting the escape tower to work. As the escape rockets ignited, they pulled the orbiter and descent modules up and away from the disaster. Behind, the launcher burst into a ball of fire. Inside the descent module, the crew shot 3,100 feet (950 m) into the air, the orbiter was jettisoned, the emergency parachute opened and they landed safely only two and a half miles (four km) away. It was the first time the escape tower had ever been used on the launch pad—and it worked.

Damage to the launchpad, however, was extensive, and it would be October 20 before the next Progress would resupply the drifting Salyut 7 and February 1984 before the next crew would arrive. Finally, eight months after the Cosmos 1443 had brought up two new solar panel extensions, Lyakhov and Alexandrov attached them to the existing panels in two EVAs of over two hours each on November 1 and 3. It was a mission originally intended for the specially trained Soyuz T-8/T-10, crews, Titov and Strekalov, who never made it. After several cold, damp weeks of discomfort due to power shortage aboard Salyut 7, the Soyuz T-9 crew completed their visitorless, event-packed mission and returned to Earth on November 23.

Getting Back Up and Running

Things went more smoothly for their successors. Arriving on February 8, 1984, aboard Soyuz T-10B, Leonid Kizim, Vladimir Solovyev and Oleg Atkov were the first three-man crew to set foot on Salyut 7 in nearly 18 months.

After resupply by Progress 19, they began to entertain visitors on April 3, with the arrival of Soyuz T-11 carrying Yuri Malyshev, Gennady Strekalov and India's first representative in space, Rakesh Sharma. The six cosmonauts spent eight days together, with the visitors returning home in Soyuz T-10B.

After another Progress resupply and extensive work on materials processing, Kizim and Solovyev settled in for a long bout of extravehicular repairs on the damaged fuel tanks and propulsion system—totalling an impressive 35 hours of EVA time between them over the coming weeks. In the entire 25-year history

of the Soviet space program only 28 hours of EVA time had previously been clocked in a total of 12 EVAs.

With the arrival of visiting Soyuz T-12 on July 17, 1984, Svetlana Savitskaya became the first woman to make a second spaceflight—once again stealing the thunder from American Sally Ride, whose second flight aboard the Space Shuttle would not occur until September 1984. On July 25, in a three-hour, 35-minute EVA—the first ever made by a woman—she one-upped U.S. astronaut Kathryn Sullivan, who was scheduled for a September EVA. With her crew mates Vladimir Dzhanibekov and Igor Volk, Savitskaya returned home aboard Soyuz T-12 on July 29 after a 12-day stay.

Just over two months later, Kizim, Solovyev and Atkov returned to Earth with a new space-endurance record of 236 days, 22 hours and 40 minutes total time from lift-off to landing, February 8 to October 2, 1984. Their record would not be broken until October 1987, when Yuri Romanenko would surpass them aboard the Mir space station, successor to Salyut 7.

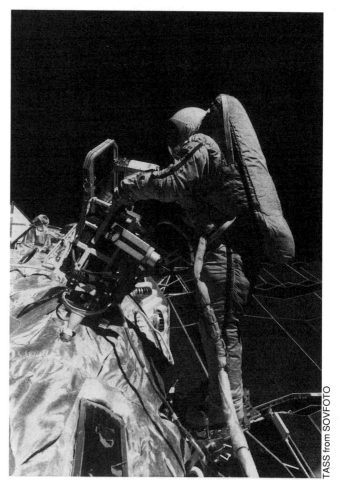

Svetlana Savitskaya becomes the first woman to perform a space walk

Soviet Space Flight Control Center during Leonid Kizim's and Vladimir Solovyov's space walk outside Salyut 7, May 1986

Their return vehicle, the Soyuz T-11, had been in space more than 180 days since its lift-off on April 3—much longer than any of the earlier series of Soyuz spacecraft. The Soviet manned space record had become truly impressive. By the end of Kizim, Solovyev and Atkov's flight the Soviet total (adding individual man-hours together) was already equal to 10 man-years in space.

Salyut 7, however, would remain unoccupied for the following six months, until June 6, 1985, and the arrival of Soyuz T-13 with Vladimir Dzhanibekov and Viktor Savinykh aboard. Theirs was primarily a fix-up mission to correct troubles with the ground-based control system and possibly a power problem. The interior, according to a reported description by Dzhanibekov, was "stuffy and cold" when they arrived.

In addition, the pair contributed agricultural photography for a project called Kursk-85—coordinating simultaneous observations by aircraft and ground crews with the Salyut 7 crew's photography, the Soviets planned to gain insights about the significance of remote-sensing data. The project helped find economical grazing areas for livestock, evaluate

pest problems in forest areas and locate underground water reservoirs.

Visited about two weeks after arrival by a routine Progress 24 supply ferry, the crew jettisoned it in July and welcomed Cosmos 1669, a laboratory module with its own electrical system. Six weeks later they sent Cosmos 1669 on its way, in preparation for the August 30 arrival of Soyuz T-14, with Vladimir Vasyutin, Georgy Grechko and Alexander Volkov on board. For the first time the six cosmonauts—essentially two resident crews at once—would work together on Salyut for nearly a month, followed by a partial exchange of crews.

On September 26, host-crew commander Dzhanibekov returned with Grechko aboard the T-13, leaving Savinykh behind to finish the mission with the Soyuz T-14 crew. It was an important new step toward the Soviet goal of a permanent presence in space, since Salyut stations had always before been left vacant for several months between long-term crews.

The next day brought yet another first: the launch of the Salyut "annex" Cosmos 1686. Docked at the Salyut 7 forward port, the Cosmos 1686 module nearly doubled the length of the space station, increasing the living/working quarters by half. The new arrangement saw only two months' use though, with complete evacuation of Salyut 7 November 21, 1985, at least in part due to illness suffered by Vasyutin, who was taken to the hospital immediately after his return to Earth. Not until spring of the following year, after the successful launch of a new breed of space station called Mir, would Salyut 7 again see human occupation. Leonid Kizim and Vladimir Solovyov, having docked their Soyuz T-15 at the new Mir station on March 15, 1986, buzzed over on May 5 to the empty Salyut 7 station (where they had spent 237 days together in 1984). They docked at the rear port and finished up some of the experiments interrupted the previous year by Vasyutin's illness.

In addition, the pair retrieved exposure samples of experimental construction materials by using a new "Girder Constructor" to deploy and retract a 49-foot x 1.6-foot (15 m x 1/2 m) truss or framework. Erecting the folded girders proved to be a speedy and efficient procedure, taking a matter of just minutes—confirming that girders of this type could possibly be used in the future to attach long-term modules to space stations.

In mid-August 1986, engines aboard the attached Cosmos 1686 were fired to boost Salyut 7 to a higher orbit 300 miles above the Earth's surface, where it should survive until 1991 to 1994. If it does, the Soviets will have demonstrated their spacecraft's capability to endure successfully in space for up to twice as long as it would take to go to Mars—and return.

Looking Back on the Salyut Experience

For the Soviets, and for space exploration in general, the Salyut experience measures large in both triumph and disappointment. From the first launch in 1971 to the present, their attempt to establish a permanent human presence in space has suffered many fitful stops and starts and great frustration. Beset by docking problems with the Soyuz craft, efforts to link up with Salyut failed four tries out of 11 between 1971 and 1977. At least two Soyuz launches were aborted, and at least two out of eight attempts to launch a Salyut station failed. Cosmonauts risked their lives in dozens of close calls, and three Salyut cosmonauts met their deaths in the unforgiving environment of space.

But no one ever said that pioneering in space would be easy, and their efforts have paid off, the setbacks offset by consistent breakthroughs and steady progress. Over the Salyut years the Soviet space program has proven that both human beings and the Soyuz craft can function for long periods in space. The longevity of their spacecraft is further shown by the fact that both Salyut 6 and 7 exceeded their designed life expectations. They've also shown that large complexes can be built from self-contained modules and orbited as a single mass. They've achieved routine resupply systems using the Progress cargo ferry, and the Soyuz T-13/Soyuz T-14 crew exchange showed that rotating crews was possible.

With the Salyut 7 missions it became clear that a permanent Soviet presence in space was now well within reach. By 1986, with the arrival of their new Mir station, the Soviets had firmly established a steadily strengthening frontier outpost—beginning with their Salyut "cabin in the sky."

PART 2

MAKING A WORKPLACE
IN SPACE

5

SATELLITES: THE PERPETUAL PRESENCE

The growth of our science and education will be enriched by new knowledge of our universe and environment, by new techniques of learning and mapping and observation...
—President John F. Kennedy
September 12, 1962

The rocket has opened up the door to space. Although humankind's natural curiosity and thirst for exploration and knowledge has managed to keep that door open, its hinges are kept oiled by more practical matters. If the rocket has made access to space possible, the satellite has made that access economically desirable.

Orbiting the Earth in a perpetual merry-go-round of activity, satellites today have become an important part of each of our lives. From our early morning television newscasts to our telephone calls, from the weather reports in our newspapers to the box scores on the sports page, from presidential decisions on affairs of state to stock market fluctuations in affairs of the economy, Earth-orbiting satellites interact with our daily lives in hundreds of ways.

Although they come in a variety of sizes and configurations, artificial Earth-orbiting satellites (artificial designating human-made, since the Moon also is an earth-orbiting satellite) are essentially objects placed in orbit above the Earth's atmosphere and which accompany the Earth in its passage through space. The "objects" may carry human passengers, as in the case of such Earth-orbiting space projects as the Soviet Salyut and Mir space stations or the American Skylab, Mercury and Gemini spacecraft. More fre-

quently they orbit with non-human payloads of various kinds. They may be "passive," carrying no active instrumentation of their own, such as the early Echo communications satellite, or they may be sophisticated packages carrying full complements of highly specialized electronic equipment.

Most artificial satellites today are of the highly sophisticated variety. A typical telecommunications satellite may carry enough advanced electronics to allow for the availability of multiple television channels, radio bands and tens of thousands of telephone circuits, all packed into a compact cylindrical-shaped body that may be no larger than 12 feet in diameter and 24 feet high.

Different Satellites for Different Jobs

Satellites can be categorized roughly by the jobs that they were designed to perform. Although hybrid satellites that do more than one job have been developed, and certain "specialized" satellites such as "Earth resources satellites" may perform non-specified "extra" jobs such as spying on neighboring nations, most satellites can be divided up into 6 major types: (1) *science satellites*, which were historically the first to reach orbit and gather scientific information about the atmosphere and near-space conditions

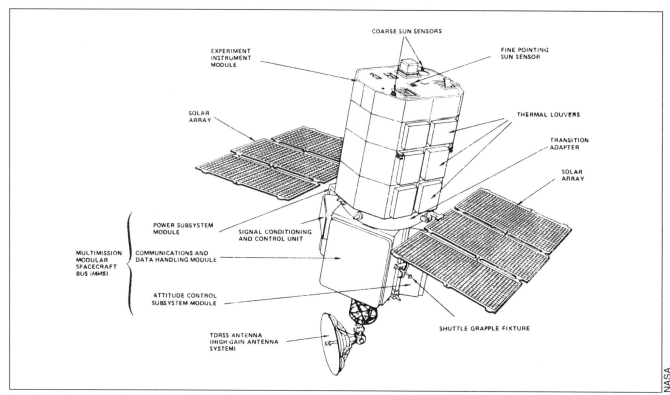

NASA

The Solar Maximum Mission scientific satellite

of the Earth; (2) *communications satellites*, the "superstars" of the commercial satellite industry; (3) *earth resources satellites*, used for land and sea study and investigation; (4) *meteorological satellites*, used for weather, science and atmospheric study; (5) *navigational satellites*, used in aiding land, air and sea navigation; and (6) *military satellites*, a "generic" label which may incorporate all or parts of each of the other types for military use.

Major Satellites: A Chronicle

10/4/57   **Sputnik:**   First artificial satellite (Soviet)

1/31/58   **Explorer 1:**   First U.S. satellite—discovered Van Allen radiation belts

3/17/58   **Vanguard 1:**   Plots Earth's shape (U.S.)

8/7/59   **Explorer 6:**   Takes TV pictures from space (U.S.)

4/1/60   **Tiros 1:**   First weather satellite (U.S.)

5/24/60   **Midas 2:**   Missile detection military satellite (U.S.)

8/10/60   **Discoverer 13:**   First recovered payload (U.S.)

8/12/60   **Echo 1:**   First telecommunications satellite (U.S.)

4/26/62   **Ariel-1/UK:**   First British satellite

7/10/62   **Telstar 1:**   First commercially financed telecommunications satellite (AT&T)

7/26/63   **Syncom 2:**   Geosynchronous communications satellite (U.S.)

8/19/64   **Syncom 3:**   Broadcasts Olympic games opening from Japan (U.S.)

12/15/64   **Marcos-1:**   First of a series of Italian satellites studying the Earth's atmosphere

4/3/65   **Snapshot 1:**   Nuclear reactor in orbit

4/6/65   **Early Bird (Intelsat 1):**   Commercial TV communications (international)

7/16/65   **Proton 1:**   Cosmic ray measurements (USSR)

10/30/67   **Cosmos 186/188:**   Demonstrates automatic docking (USSR)

10/20/68   **Cosmos 249:**   Inspection/destruction satellite (USSR)

2/11/70   **Osumi:**   First Japanese satellite

4/24/70   **China 1:**   First Chinese satellite

7/23/72   **Landsat 1:**   First Earth resources satellite (U.S.)

4/19/75   **Aryabhata:**   First Indian satellite

11/23/77   **Meteosat:**   European weather satellite

6/27/78   **Seasat 1:**   Sea resources satellite (U.S.)

5/17/79   **Cosmos 1099:**   First of new series of Earth resources satellites (USSR)

9/25/79   **Cosmos 1129:**   First biosatellite, international experiment to breed mammals in space (USSR)

12/6/80   **Intelsat 5:**   Biggest communications satellite to date (international, 105 nations)

4/8/84   **STW/China:**   First Chinese telecommunications satellite in geosynchronous orbit

8/4/84   **Telecom-1:**   French telecommunications satellite

2/8/85   **Arabsat:**   Telecommunications satellite for Arab countries, launched by Ariane 3

2/8/85   **Brasilsat-1:**   Brazilian telecommunications satellite, launched by Ariane 3 rocket

6/17/85   **Morelos:**   First Mexican telecommunications satellite, launched by the U.S. Shuttle

8/27/85   **Aussat:**   First of a three-satellite Australian telecommunications system, launched by U.S. Space Shuttle

2/22/86   **Viking:**   Swedish scientific satellite for studying the interaction of hot and cold plasma

2/22/86   **SPOT 1:**   French remote-sensing satellite (Earth resources)

3/89   **ERS-1:**   Earth resources satellite (ESA)

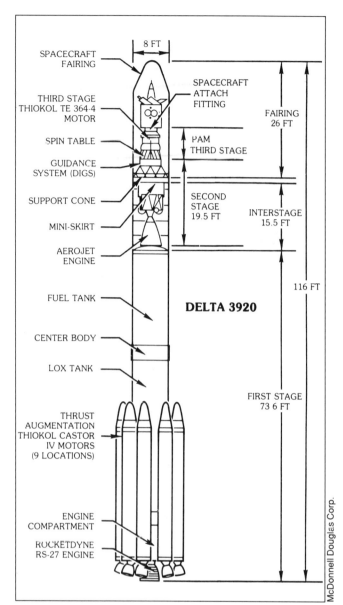

SPACECRAFT
FAIRING

8 FT

SPACECRAFT
ATTACH
FITTING

THIRD STAGE
THIOKOL TE 364-4
MOTOR

FAIRING
26 FT

SPIN TABLE

PAM
THIRD STAGE

GUIDANCE
SYSTEM (DIGS)

SUPPORT CONE

SECOND
STAGE
19.5 FT

INTERSTAGE
15.5 FT

MINI-SKIRT

AEROJET
ENGINE

116 FT

FUEL TANK

DELTA 3920

CENTER BODY

LOX TANK

FIRST STAGE
73 6 FT

THRUST
AUGMENTATION
THIOKOL CASTOR
IV MOTORS
(9 LOCATIONS)

ENGINE
COMPARTMENT

ROCKETDYNE
RS-27 ENGINE

McDonnell Douglas Corp.

Delta expendable launchers have successfully launched
more than 150 payloads into orbit, including Echo, Telstar 1
and Early Bird communications satellites

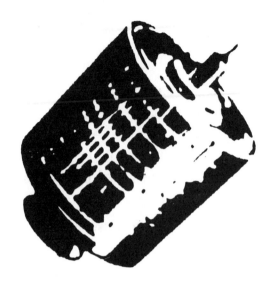

NASA

An Applications Technology Satellite—one of a series of
communications, meteorological and navigation satellites.
ATS-6 was launched in an educational project to beam TV
programs to village centers in underdeveloped countries

NASA

Astronaut Dale Gardner prepares to rescue a spinning Wes-
tar communications satellite during U.S. Shuttle mission 51- A

The Beginning: The Scientific Satellites

On October 4, 1957, the Soviets took the world by
surprise when they orbited the first successful artificial satellite. Nicknamed Sputnik, the tiny, 22.8-inch
diameter (58-cm), 184.3-pound sphere carried a min-
imum of scientific equipment designed merely to
measure temperatures and atmospheric densities
briefly. Its impact, though, was far heavier than its
payload. With Sputnik the world suddenly had
entered the space age.

The United States quickly followed suit. After a couple of embarrassing failures, America successfully orbited its first satellite, Explorer 1, on January 31, 1958. With a total weight of only 30.8 pounds it was 6 feet 8 inches in length. Explorer 1 gave the United States its first scientific returns by satellite when it confirmed the existence of the Van Allen radiation belt that surrounds the Earth and helps protect our planet from cosmic rays.

By the mid-1960s the USSR had orbited a score of unmanned satellites in its Cosmos series and the United States after numerous failures and only a few successes in its Vanguard series, was regaining space respectability with its continuing successful and long-running Explorer series of scientific research satellites.

In December 1968 the United States orbited its first large-scale astronomical observatory OAO 2 (Orbiting Astronomical Observatory 2), nicknamed Stargazer, and in August 1972 it launched the more sophisticated OAO 3 called Copernicus.

Science had entered the satellite age, and by the late 1970s the United States, the USSR, Great Britain, West Germany, Italy, Japan and China all had dedicated science satellites in orbit.

Science had entered the satellite age, and by the late 1970s the United States, the USSR, Great Britain, West Germany, Italy, Japan and China all had dedicated science satellites in orbit.

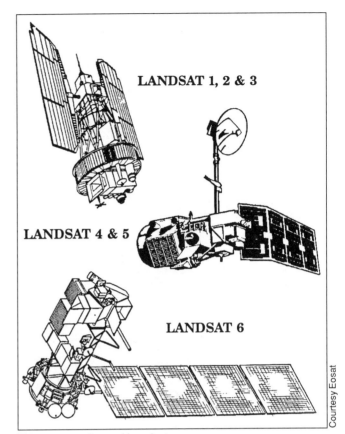

Three generations of Landsats

But while science had led the way into space, practical applications were not far behind.

An Echo of Things to Come: The Communications Revolution

The age of communications satellites, so much a part of our everyday world and the prime commercial focus of satellite technology, began quietly enough with the launch of NASA's ECHO 1 on August 12, 1960.

A passive satellite, carrying no transmitting instruments of its own, Echo 1 was simply a 98.42-foot balloon with a high-tensile strength Mylar polyester coating which, when inflated in space, reflected 98 percent of the radio waves aimed at it. Bounced back toward various receiving stations on Earth, the reflected signals of ECHO 1 first carried a coast-to-coast message from President Dwight Eisenhower on the day of its launch. It wasn't fancy but it worked. For the first time human beings around the world heard a voice reflected back from space. More wonders were literally waiting in the wings.

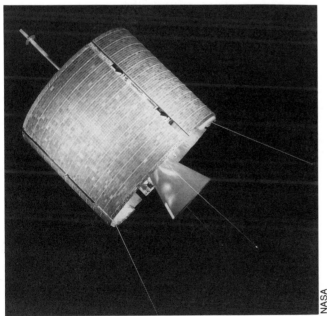

Artist's concept of an Early Bird satellite in flight

Landsat 4 view of New York City

The world really began to take notice of the communications revolution with the deployment of Syncom 3. It was launched by NASA in August 1964 into geostationary orbit (with a period equal to Earth's rotation—23 hours, 56 minutes, 4.1 seconds) so it seemed to hover in one spot, above the International Date Line, an arbitrary north-south line in the Pacific Ocean, west of which is a day later, by international agreement, than it is to its east. From there, Syncom 3 allowed viewers worldwide to watch a broadcast of the opening ceremonies of the Tokyo Olympics.

By April 1965, a little less than eight years after the first artificial satellite caught the world's attention, Intelsat 1, nicknamed "Early Bird," became the first commercial geostationary communications satellite. Launched by NASA for the international organization INTELSAT (International Telecommunications Satellite Consortium), Early Bird brought the entire world into the communications revolution. Despite its initial modest capacity (only 240 telephone circuits or one television channel), with its launch the marriage of space to the communications industry became big business. Today over 125 communications satellites from many different nations and consortiums circle the globe and a constant flurry of activity surrounds them, from launches to repairs, upgrades and replacements. They demonstrate daily the commercial benefits of space and guarantee by their highly profitable presence that the space age is here to stay.

The Earth Watchers: The Landsat Revolution

While not achieving the same superstar status as communications satellites, Earth resources satellites have become valuable assets themselves in an increasingly environmental- and resource-conscious world. Perhaps best represented by NASA's Landsat series, these satellites scan the Earth from orbit with sophisticated "remote-sensing devices," including high-resolution cameras (which can take "close-ups" from orbit) and infrared sensors (which measure invisible heat waves just beyond the red end of the visible spectrum). Earth resources satellites have spotted growth patterns of vegetation in the jungles, fish movement in the high seas, likely mineralogical deposits in the mountains and ancient lake beds and rivers in the desert. Whether monitoring the insidious spread of crop disease or pinpointing the most beneficial areas for planting, Earth resources satellites have made the thoughtful management of land and sea a possibility never before achievable on such a wide scale.

Telstar 1, the world's first privately financed communications satellite, was launched by NASA on July 19, 1962. Capable of carrying 600 telephone channels or one television channel, its low orbit, however, allowed transmissions to be made for only 20-minute periods.

Relay 1, launched by NASA in December of that same year, was an experimental communications satellite built by RCA for NASA. Put into a higher orbit, its transmission times were longer than Telstar's.

Using the Spectrum:
Astronomy's Many Ways of Looking at the Universe

When we look at the Moon, planets and stars, even with binoculars or a telescope, we're using only one of the many ways astronomers "look" at the universe. Scientists have discovered a wide variety of ways to learn about these fascinating but distant objects. When we see light coming from the Sun, Moon or stars, for instance, we're observing only one area—visible light—of the spectrum (or range) of electromagnetic radiation.

However, most objects in the universe emit electromagnetic radiation (waves or rays created by variations in magnetic and electric fields) of several kinds—not just the rainbow of colors we see in visible light. Beyond the red end of the visible spectrum, the longer wavelengths and slower frequencies of infrared radiation and radio waves can be detected by instruments such as infrared telescopes and radiotelescopes. Beyond the violet end of the visible spectrum, the short, high- frequency waves of ultraviolet radiation, X-rays and gamma rays can be detected by special instruments such as ultraviolet spectrometers, X-ray telescopes and gamma ray telescopes.

Satellites and spacecraft such as Salyut, Skylab, the Shuttle and Mir provide ideal platforms for observing the objects of the universe in all ranges of the spectrum—because reception is unobstructed by the atmosphere, pollution and various conflicting signals from Earth.

Many nations are also now recognizing the usefulness of turning some of these and other remote-sensing instruments toward the Earth from space to "take the vital signs" of the planet—to detect patterns of change in the atmosphere, in living things such as the tropical rain forests, in the oceans and in the Earth's crust.

Launched by NASA in July 1972, Landsat 1, the first of the Landsat series, sent back over 300,000 detailed pictures of the Earth in its nearly six-year lifetime. Continuing throughout the present and despite some political and financial wranglings over its future, the Landsat series, along with its counterparts such as the ocean observing Seasat and the French Spot satellite, continues to be an effective tool in the process of mapping and managing the Earth's divergent and fluctuating supply of valuable resources. An even more advanced member of the series, Landsat 7 is currently scheduled for launch in the late 1990s.

Space-Age Weather Watchers: Meteorological Satellites

Beginning with Tiros 1, launched by NASA in April 1960, meterological satellites have continued to monitor the Earth's environment and atmosphere.

The USSR launched its first experimental meteorological satellite Cosmos 23 in December 1963. In the intervening years a steady stream of data from more advanced American and Soviet systems has allowed people to monitor and understand the Earth's complex weather environment on a level impossible before the space age. From their orbital vantage points, meterological satellites can observe weather conditions and patterns in areas of the globe beyond access to conventional Earth-bound means. In this way they make possible a much larger-scale portrait of the complicated interactions between such weather systems. Although there is still much to learn about the forces that drive the Earth's weather, meteorological satellites act as advanced early warning systems for such impending disasters as hurricanes and crop-damaging weather changes, and they advance our knowledge and understanding of the Earth's atmospheric environment daily.

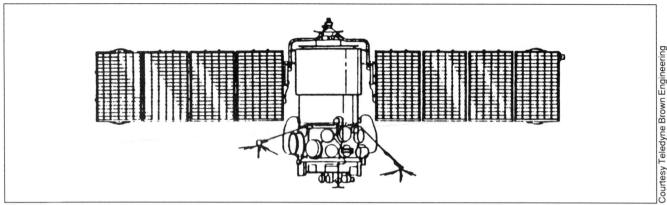

Courtesy Teledyne Brown Engineering

Meteor 2, a Soviet weather satellite

Navigating by the Artificial Stars: Navigation and Search and Rescue Satellites

First developed to aid submarines in missile launches against enemy targets, navigation satellites have since become an effective tool for military and non-military surface ships, land and air vehicles and search and rescue missions. The first in a series of American navigational satellites, Transit 1A was launched in September 1959 but failed to reach orbit. Transit 1B, launched in April 1960, achieved orbit but not the one needed for its purpose. After a difficult history of launch and orbital failures, Transit 5A1 became the first truly operational navigation satellite shortly after its launch in December 1962.

The USSR, after a similar shaky start, launched its first operational navigation satellite Cosmos-192 in November 1967. In February 1978, the United States began to launch its more sophisticated Navstar series with the success of Navstar 1. The Soviet Union began launching its more advanced Glonass series with three satellites—Cosmos 1413, Cosmos 1414 and Cosmos 1415—in October 1982.

In 1982 the international satellite-based search and rescue system SARSAT-COSPAS became operational. The program involved international cooperation between Canada, France, the Soviet Union and the United States and since its inception has saved over 500 lives with its ability to quickly trace the movements of land, sea and air vehicles that are in trouble.

Strategic Space: Military Satellites

The militarization of space, by now involving the presence of literally hundreds of military satellites in the skies, has been in effect since the early 1960s.

(Some 2,000 have been launched by the United States alone in less than 30 years.) The satellite types fit roughly into the same categories as the satellites already mentioned—but the application is military, the data acquired is secret, and often the technology may be more sophisticated. They fall into four groups: reconnaissance satellites, intelligence satellites, communications satellites and navigation satellites. The first two types are used in gathering data, while the last two are used within the military to aid in communication, deployment, navigation and so on.

More sophisticated cousins of Landsat, reconnaissance ("spy") satellites make use of high-resolution cameras to collect information about sites of military interest. Reputed to be able to photograph objects even just a few inches in size, spy satellites are, of course, top secret.

The exact capabilities of intelligence satellites, such as the one carried to orbit by the U.S. Shuttle in early 1985, also remain heavily guarded, but the existence of sensitive "electronic snoopers," used in part to intercept radio communications, is indisputable. The Soviets have launched four intelligence satellites over the past decade, including one in 1984, Cosmos 1,603, that appears to be the largest and most complex to date. The United States used its second Shuttle launch after the long delay following the Challenger tragedy, in December 1988, to make its own high tech addition to its many "spies in the sky."

The U.S. Navstar Global Positioning System, expected to be functioning by the late 1980s, is an example of a military satellite navigation system. Composed of 18 satellites in three orbital planes, the system will provide precise navigational data to the Department of Defense on a coded channel. Location and speed of land, air or sea vehicles can be reported by the system with precision. It will also be able to detect atmospheric tests of nuclear weapons anywhere on our planet.

For designers of military communications satellites the challenge is to keep communications from being intercepted, and the level of sophistication is constantly increasing. In the United States, the military relies on satellites for more than two-thirds of its long-range communications, making invulnerability essential.

With the rise in importance of military satellites, of course, anti-satellite weapons become a key concern in defense. A simple intercontinental ballistic missile (ICBM) can destroy a satellite that a nation is relying on for vital information. One way the military combats this vulnerability is by having many backup satellites for every function—partially explaining the large number of military satellites in our skies.

Regardless of their specific purpose, the presence of satellites of all kinds in the skies has served to make our world smaller and has made humanity more aware of our interconnectedness. A communications satellite makes a relative in Cologne or Australia just a phone call away. The ravages of poor crop management by the early Maya in Mexico are spotted by researchers poring over Landsat images. A fishing

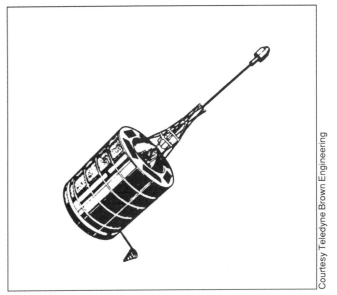

A Soviet navigation satellite

boat lost in a squall finds its way home. The Earth no longer seems so vast.

Or so limitless.

Leasat 4 satellite being released from the shuttle bay. It is a communications satellite leased by the U.S. Department of Defense for UHF communications among ships, planes and fixed facilities

6

A TROLLEY FOR SPACE: THE U.S. SPACE SHUTTLE

I think man is needed in space. You can't do the things we have done up here with unmanned space probes. You need brains, you need minds up here that can think, that are innovative.
—Senator Jake Garn
STS-51D Payload specialist

I think that science is the stuff that pays for itself on these missions. It's going to improve the quality of life down here.
—Frederick Gregory
Pilot, Shuttle Mission STS-51B

As the high-profile Apollo Moon missions ended and, after them, the Skylab Project closed down, the American space community and NASA turned their attention toward what was viewed by many as the next logical step in the conquest of space: The idea of a winged recoverable and reusable spacecraft—a concept that had been kicking around since the early 1950s. The perfect spacecraft, in the minds of many, would take off and fly through the atmosphere like an airplane, leave the atmosphere and enter space like a rocket, and then return to Earth and land like an airplane again.

The dream and the reality, though, were far apart, and the separation between them was not only technological but social and economical.

The Apollo Project had been successful and enormously expensive. Americans had walked on the Moon, but young American soldiers were now walking in increasing numbers through the jungles of Vietnam and in the growing political and social unrest of the late 1960s the bloom was definitely fading from the space rose. In view of the political realities, NASA needed a large project to keep itself alive and the space era open, but the American public and politicians were becoming increasingly jaded and wary about backing such projects. To make matters worse, as space funding began to dry up, the space science community itself became increasingly divided, with one camp pushing for continued "manned" missions and a smaller but increasingly vocal opposition pushing for more "unmanned" missions such as the successful Viking Project to Mars and the stunning Voyager missions to the outer planets—Jupiter, Saturn and Uranus.

U.S. Space Shuttle—A "Space Transportation System"

Orbiters: Columbia, Challenger (destroyed in explosion, January 28, 1986), Discovery, Atlantis

Orbiter Size: *Length*: 122 feet 2.5 inches (37.25 m)

Wingspan: 78 feet 0.07 inches (23.78 m)

Tail Height: 56 feet 7 inches (17.25 m)

Orbiter Weight: (fleet average) 149,466 lbs (67,797 kg)

Crew Capacity: Eight (first carried on STS-61A in 1985, the 22nd Shuttle flight)

Cargo Bay Capacity: 15 by 60 feet (4.57 by 18.3)

First Test Flight: STS 1, April 12, 1981 (Columbia)

Description: What NASA calls the Space Transportation System (STS) is a combined system consisting of several parts: rocket boosters for takeoff (two expendable solid-fuel rockets plus a reusable external tank (ET) that carries liquid fuel) and the orbiter itself—which is what many people think of as the Space Shuttle.

The orbiter is able to land on a special runway by gliding in from orbit either piloted or on automatic pilot. Its cargo bay can be adapted to contain pressurized modules such as ESA's Spacelab (see box) or it can hold up to three satellites at one time. It has been used to launch, retrieve and repair satellites, as well as for numerous experiments in space.

The Shuttle cannot, however, take off from an airstrip on its own. So, because launches generally take place from the East Coast of the United States, at Kennedy Space Center in Florida, and the Shuttle usually lands at Edwards Air Force Base in California's Mojave Desert on the West Coast, the orbiter returns to the East Coast piggyback atop a Boeing 747.

Problems: Numerous difficulties—loosened tiles on reentry, delayed launches, tight scheduling, spotty funding from Congress and so on—have plagued the Shuttle. But its greatest setback occurred on January 28, 1986, when the Challenger, STS 51-L, blew up before a television audience of millions. The crew of seven included Christa McAuliffe, who was to be America's first teacher in space. After an arduous two and a half years of redesign and testing, however, the Shuttle is flying once again.

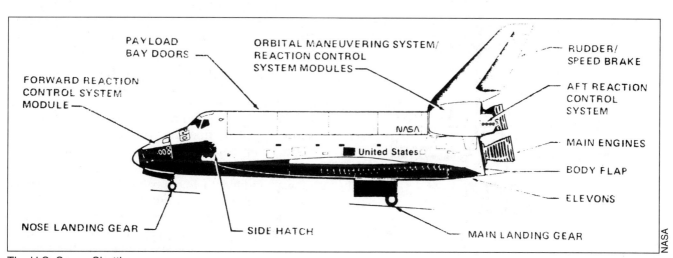

The U.S. Space Shuttle

Into such strife the United States Space Shuttle was born. Small wonder that its birth was hard and its young life blighted.

Sitting on its launchpad at Kennedy Space Center in Florida on April 12, 1981, STS-1 was a sight to behold. The initials stood for NASA's rather unglamorous designation of the project—Space Transportation System—but to the public it was the Space Shuttle and the maiden flight of the first Shuttle orbiter, Columbia. Twenty years after Yuri Gagarin's historic first spaceflight, and six years after the last American had flown in space in the Apollo-Soyuz Test Project, a new era of human spaceflight was about to begin.

After years of financial and political wrangling, budget cutting, cost-overruns, design changes, engineering and technical difficulties, engine problems and heat shield problems, the Shuttle program was about to fly.

The "flying machine" itself was less than beautiful.

What the more than a million spectators saw who lined the Florida beaches and watched on television that day was an ungainly configuration. STS-1 looked more like it had been tinkered together by a back-room inventor than the highly sophisticated product of one of the world's two leading space contenders.

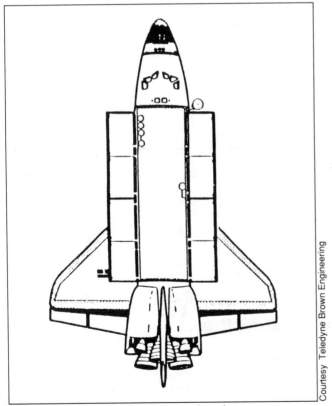

Top-down view of Shuttle with cargo bay doors open

Courtesy Teledyne Brown Engineering

Gone, at least temporarily, was the idea of a sleek "rocket-plane" that would take off from a runway under its own power and fly to space. Columbia, the first of four Shuttle orbiters as the spacecraft were now called, perched instead on the back of an assemblage that consisted of a giant external fuel tank and two solid rocket boosters. Before being jettisoned in separate stages, the external tank and boosters would lift the spacecraft from its launchpad and into orbit. Flying in orbit, it was capable of highly complicated maneuvers, but once returned to the Earth's atmosphere, the orbiter would awkwardly and ingloriously turn into the world's heaviest and fastest-falling glider as it headed for its landing site at Edwards Air Force Base in the hot, dry California desert.

A "flying brickyard," one astronaut later called it, but it was the world's most advanced, computerized and complex flying machine.

Manning the first flight were flight commander John Young and pilot Robert Crippen. Young, age 50, was an experienced space program veteran who had flown four previous space flights in the Gemini and Apollo programs, including the Apollo 16 lunar landing and a three days' stay on the Moon's surface. Crippen, age 43, had waited over 14 years to fly in space. Coming into NASA from the USAF Manned Orbiting Laboratory program (a planned platform in space that never got off the ground), he had been too late for Apollo or Skylab.

At 7:00 A.M. (EST) America's first space shuttle lifted off of Pad 39A at Kennedy Space Center and began its climb into orbit. Approximately 12 minutes later it arrived. For the first time in nearly six years Americans were in space again, and a new era of space exploitation had begun.

With the Space Transportation System, NASA had argued, space trips could be made on a routine basis. The idea of a "shuttle" ferrying payloads of astronauts, scientists, commercial satellites—and eventually the parts and pieces that would go into the assembling of America's first full-scale dedicated space station—was certainly an appealing one, but the Space Shuttle had some strikes against it even before it had its first time at bat. Cost-cutting and design compromises had brought it "on-line" behind schedule and woefully short of its initial promise.

It was limited to low-Earth orbit, although most communications and commercial satellites had to be placed in the higher geosynchronous orbit, which meant satellites requiring higher orbit required additional booster rocket assistance. The Shuttle's stay in orbit was limited to around 10 days, depending upon the size of the crew, and its turnaround time on Earth was complicated by the fact that it couldn't even fly on its own from its landing site back to its launch site, but had to be carried like a baby, piggyback, on top of a Boeing 747.

Still, two days later, on April 14, 1981, as Young and Crippen guided the shuttle Columbia home from orbit to its landing site in the Mojave Desert, NASA had much to be proud of. Columbia's first mission— primarily to test out the orbiter's systems—had gone smoothly. Touching down on the desert runway at the speed of nearly 215 miles an hour, twice as fast as commercial airliners ordinarily land, STS-1 had been a resounding success.

Satellites Delivered or Repaired on the Spot!

NASA was in the space shuttle business. For a while it looked like the dreams of the early space pioneers

were on their way to being realized. Americans now could commute to space.

From November 12, 1981, through July 4, 1982, the United States launched three more Shuttle missions flying the Columbia orbiter, STS-2, STS-3 and STS-4, all primarily designed to continue testing the Shuttle's systems.

On November 11, 1982, STS-5 (Columbia) made its first fully "operational" flight. With Vance Brand as commander and Robert Overmyer as pilot, STS-5 also carried aboard Joseph Allen and William Lenoir as mission specialists. The primary mission in this case was to deploy two communications satellites, the first to be launched by the U.S. Space Shuttle. While in orbit the crew also ran through a series of scientific and engineering experiments.

Launched at 7:19 A.M. (EST) Columbia quickly reached orbit and spent five days, two hours and 14 minutes in space while orbiting the Earth 81 times before returning to land at Edwards Air Force Base in

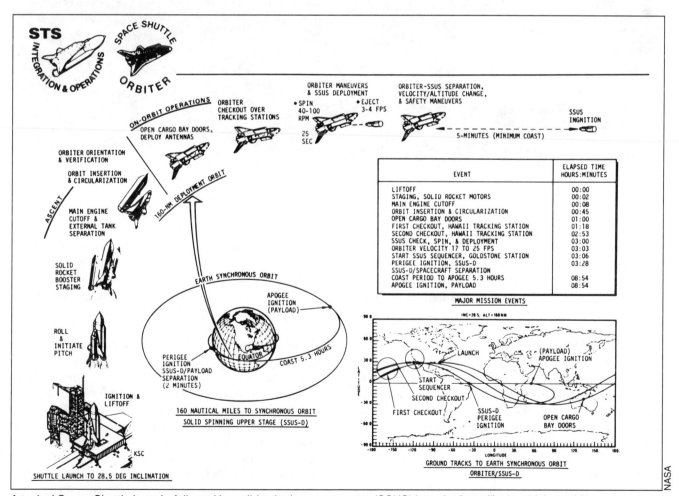

A typical Space Shuttle launch, followed by solid spinning upper stage (SSUS) launch of satellite into higher orbit

NASA

STS-5 crew members Vance Brand (holding sign), and (clockwise from upper left) William Lenoir, Robert Overmyer and Joseph Allen after successfully deploying two communications satellites

California on November 16, 1982. The flight went off without any major difficulties, although a scheduled EVA had to be canceled because of problems with a space suit.

With mission STS-6 (April 4, 1983-April 9, 1983) NASA unveiled its second orbiter. After a two-month delay to replace a defective main engine, the Challenger orbiter lifted off from the Kennedy Space Center at 1:30 P.M. (EST) on April 4, 1983. Commanding the mission was Paul Weitz, a 50-year-old retired Navy pilot, and piloting the mission was Karol Bobko. Also aboard were mission specialists Donald Peterson and Story Musgrave. The primary assignment of STS-6 was the deployment of a Tracking and Data Relay Satellite, the first in a satellite communication system (TDRSS) which would provide coverage from geosynchronous orbit for all Space Shuttle flights. Although the satellite was deployed successfully on

the first day of the five-day mission, a malfunction of the inertial upper-stage rocket, which was to have carried it to its higher orbit left it orbiting short of the altitude needed for its function. It was a problem that wouldn't have occurred if the Shuttle had been able to soar directly to the necessary geosynchronous orbit with the satellite on board, and it would take a highly skilled ground team over 59 days to carefully fire the satellite's six tiny engines through a highly complex sequence and raise it to its proper height. Still, though, Challenger itself performed well and Musgrave and Peterson became the first Americans in nine years to space walk when they spent over four hours in the mission's scheduled EVA.

STS-6 returned to Earth and Edwards Air Force Base at 10:53 A.M. (PST) after a mission that lasted five days, 24 minutes and 32 seconds, orbiting the earth 80 times.

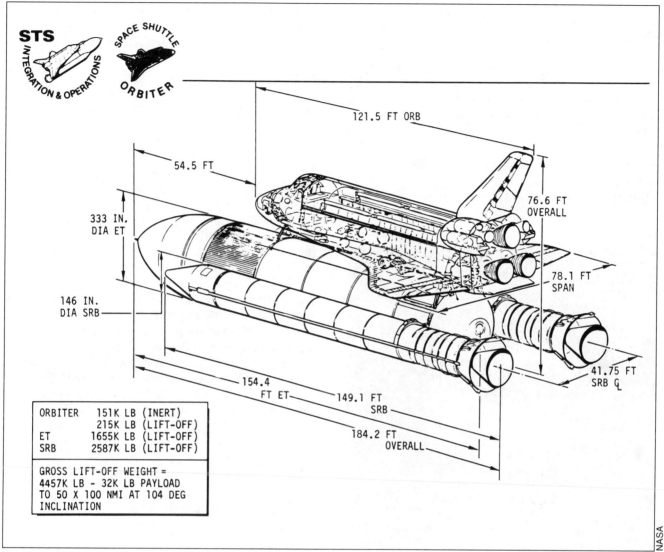

STS
INTEGRATION & OPERATIONS

SPACE SHUTTLE
ORBITER

121.5 FT ORB

54.5 FT

76.6 FT
OVERALL

333 IN.
DIA ET

78.1 FT
SPAN

146 IN.
DIA SRB

41.75 FT
SRB ₵

154.4
FT ET

149.1 FT
SRB

184.2 FT
OVERALL

ORBITER	151K LB (INERT)
	215K LB (LIFT-OFF)
ET	1655K LB (LIFT-OFF)
SRB	2587K LB (LIFT-OFF)

GROSS LIFT-OFF WEIGHT =
4457K LB - 32K LB PAYLOAD
TO 50 X 100 NMI AT 104 DEG
INCLINATION

NASA

The Space Shuttle attached to external tank and solid fuel boosters

Returning piggyback by way of Boeing 747 to Florida, Challenger's next mission, STS-7, marked another significant first. For the first time in the history of the U.S. space program, America flew a woman in space. Launched June 18, 1983, at 7:33 A.M. (EST) STS-7 carried a five-person crew. Flying with physicist Sally Ride, the nation's first "space woman," were two other mission specialists, John Fabian and Norman Thagard. Robert Crippen commanded the mission and Frederick Hauck was the orbiter's pilot. The mission deployed two communications satellites. The crew also deployed and then retrieved an experimental science "pallet" using the Shuttle's new remote manipulator arm, demonstrating the Shuttle's ability not only to deliver, but to pick up space objects.

STS-7 was to have performed another first by landing near its launch site at the Kennedy Space Center in Florida, but weather conditions were uncooperative and the orbiter put its wheels down at Edwards in California on June 24, 1983, at 6:57 A.M. (PDT), six days, two hours, 24 minutes and 10 seconds after lift-off. It had made 97 Earth orbits.

Challenger continued its forays into the heavens with mission STS-8, the Shuttle program's first night launch and landing (August 30, 1983 2:32 A.M. [EST] to September 5, 1983, 12:41 A.M. [PDT]. Commanding was Richard Truly with Daniel Brandenstein as pilot. Dale Gardner, Guion Bluford (the first Black American in space) and William Thornton served as mission specialists.

Science: And Science-Fiction Becomes Reality

The Shuttle was proving its stuff, but it had been designed for far more than basic experimentation and delivery and retrieval of satellites. Mission STS-9 (November 28, 1983 to December 8, 1983), which returned the Columbia orbiter to service, was the Shuttle's first dedicated science mission using the jointly sponsored European Space Agency/NASA Spacelab. Fitting into the Columbia's cargo bay, Spacelab 1, a completely outfitted miniature laboratory, ran a full complement of space science

Bruce McCandless, first "cherry picker" in space, using the mobile foot restraint on Canada's remote manipulator arm

experiments and observations as verification of its capabilities. Commanded by veteran astronaut and Moon-walker John Young, STS-9 was piloted by Brewster Shaw and carried two mission specialists, Robert Parker and Skylab veteran Owen Garriott, as well as two payload specialists, Byron Lichtenberg and Ulf Merbold, the first non-astronaut scientists to fly in the Shuttle program.

The Shuttle's next mission, STS-41B (using NASA's new shuttle designation system), offered a spectacular space walk. Flying the Challenger orbiter commanded by Vance Brand and piloted by Robert Gibson, with mission specialists Bruce McCandless, Ronald McNair and Robert Stewart aboard, the mission was launched February 3, 1984, at 8:00 A.M. (EST) from Kennedy Space Center. In orbit, astronauts McCandless and Stewart used the Manned Maneuvering Units, strap-on-the-back devices that looked like gizmos from the old science-fiction movies, for flying untethered outside the spacecraft. Besides testing the MMU's, STS-41B deployed two communications satellites, which failed to reach proper orbit, and also released the refurbished West German shuttle pallet satellite (SPAS). (Originally flown on STS-7, this platform loaded with scientific instruments was the first satellite ever to be flown, retrieved, refurbished and flown again.) STS-41B spent six days, 23 hours and 40 minutes in space and returned to Earth and Edwards Air Force Base on February 11, 1984, at 7:17 A.M. (EST) after making 127 Earth orbits.

The Story of "Solar Max"

The definition of "living and working in space" was extended with the Shuttle's next mission. NASA had lauded the Shuttle's ability to permit instant repairs of defective satellites in orbit, and in the cost-cutting climate of the 1980s much of the "man in space" debate had come down to the argument that human beings could make "on-the-spot" instant decisions that machines were incapable of. Retrieving, examining, and repairing a satellite in orbit would go a long way toward proving the point, not only in justification of the Shuttle program, but to the threatened philosophy of "keeping a human presence" in space as well.

The job set for the STS-41C crew was to take the Challenger orbiter into orbit and retrieve, repair and redeploy the Solar Maximum Mission satellite that had been launched by conventional rocket four years before. The satellite, nicknamed "Solar Max," had been launched as a Sun-watching observatory but had stopped functioning properly after its first 10 months of operation.

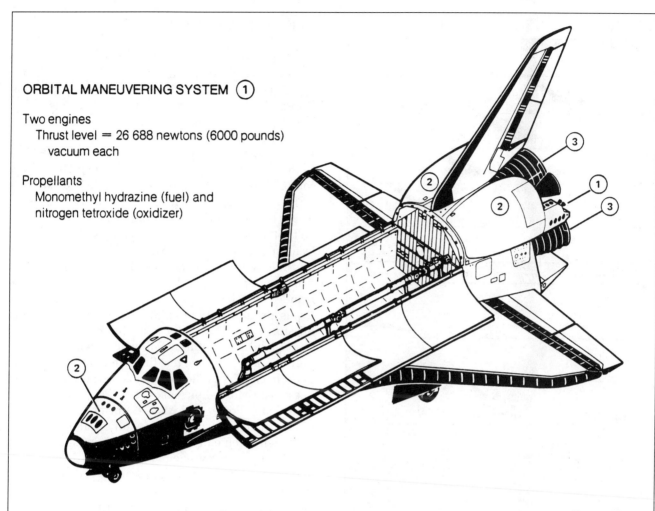

ORBITAL MANEUVERING SYSTEM ①

Two engines
  Thrust level = 26 688 newtons (6000 pounds)
    vacuum each

Propellants
  Monomethyl hydrazine (fuel) and
  nitrogen tetroxide (oxidizer)

REACTION CONTROL SYSTEM ②

One forward module, two aft pods

38 primary thrusters (14 forward, 12 per aft pod)
  Thrust level = 3870 newtons (870 pounds)

Six vernier thrusters (two forward, four aft)
  Thrust level = 111.2 newtons (25 pounds)

Propellants

  Monomethyl hydrazine (fuel) and
  nitrogen tetroxide (oxidizer)

MAIN PROPULSION (See section 2) ③

Three engines
  Thrust level = 2 100 000 newtons (470 000 pounds)
    vacuum each ·

Propellants

  Liquid hydrogen (fuel) and
  liquid oxygen (oxidizer)

NASA

The Shuttle propulsion systems

The crew for STS-41C Challenger included Robert Crippen as commander and Francis Scobee as pilot. George Nelson, James Van Hoften and Terry Hart flew as mission specialists.

Launched from the Kennedy Space Center April 6, 1984, at 8:58 A.M. (EST), Challenger reached the highest altitude yet achieved by a shuttle (310 miles above the Earth, compared to the usual 150 to 200 miles) and maneuvered to 300 feet from the malfunctioning Solar Max.

After some difficulty stabilizing the spinning satellite the crew was able to capture it, make the necessary

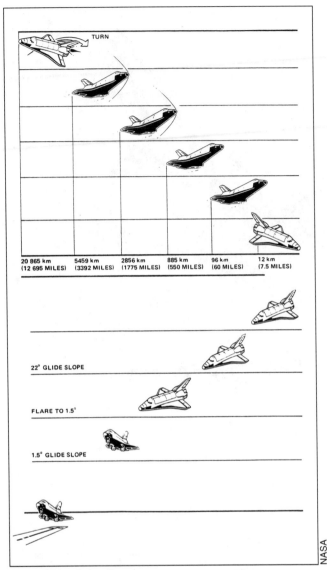

TURN

20 865 km (12 695 MILES)   5459 km (3392 MILES)   2856 km (1775 MILES)   885 km (550 MILES)   96 km (60 MILES)   12 km (7.5 MILES)

22° GLIDE SLOPE

FLARE TO 1.5°

1.5° GLIDE SLOPE

NASA

Typical Shuttle landing sequence

repairs using the free-flying Manned Maneuvering Units (MMUs) for their extensive space-walking activities, and then return it to orbit.

During the six-day, 23-hour, 40-minute mission the crew also deployed the Long Duration Exposure Facility, a platform holding 57 experiments to be left in space and retrieved by a later Shuttle mission.

Returning to Earth on April 13, 1984, at 5:38 A.M. (PST) STS-41C had made 108 earth orbits and proved that humans could make on-the-spot, in-orbit repairs of space satellites.

Things were looking good for the Shuttle program; NASA was walking around in a brand new suit. But the material had been stretched too far, and the seams were beginning to come loose.

Discovery Comes On Line

The third orbiter in the Shuttle fleet, Discovery, was to fly its maiden mission STS-41D on June 25, 1984. Three major delays involving the orbiter's engines and computers (the first discovered only four seconds before lift-off), however, postponed the launch until August 30, 1984. Once in orbit, everything went smoothly enough for Discovery and its crew of six; three satellites were deployed successfully and other experiments accomplished, and the spacecraft returned to Edwards AFB September 5, 1984, at 6:37 A.M. But the early delays had left an uneasy feeling in the air.

STS-41G found Challenger flying again. Launched October 5, 1984, at 7:03 A.M. (EDT) from Kennedy, the mission with its seven members was the largest American crew yet to fly in space at one time. Commanded by Robert Crippen and piloted by Jon McBride, it was also the first mission to fly two women aboard, mission specialists Sally Ride (making her second spaceflight) and Kathryn Sullivan (making her first). Sullivan also became the first American woman to perform an EVA as she joined mission specialist David Leestma in a three-hour space walk. Payload specialists Paul Scully-Power and Marc Garneau completed the Challenger crew, which successfully demonstrated the first satellite refueling in space—replacing spent fuel for the satellite's rocket engines. They also deployed yet another satellite along with its science and engineering duties.

Landing at Kennedy Space Center on October 13, 1984, at 12:26 P.M. (EDT) STS-41G had spent eight days, five hours and 23 minutes in space, making 133 orbits.

Things definitely seemed to be getting back on track and mission STS-51A launched November 8, 1984, and returned to Earth on November 16, 1984, seven days, 23 hours and 45 minutes later, seemed to reaffirm that view. Commanded by Frederick Hauck and piloted by David Walker with mission specialists Anna Fisher, Dale Gardner and Joseph Allen aboard, the Discovery orbiter deployed two communications satellites and retrieved and returned to Earth two others that had been deployed during mission 41B in February 1984.

The success of the Shuttle's next mission, STS-51C, is an open question. A top-secret military mission for the Department of Defense, it was launched aboard the Discovery orbiter on January 24, 1985, at 2:50 P.M. (EST) and returned to KSC three days, one hour, and 33 minutes later on January 27, 1985, at 4:23 P.M. (EST).

STS-51D and STS-51B, launched on April 12, 1985 and April 29, 1985, respectively (using the Discovery and Challenger orbiters), carried a U.S. senator, Edwin Jacob "Jake" Garn of Utah, on the April 12 Discovery mission, and two monkeys and 24 rats on the Challenger mission of the 29th. Both went smoothly enough and STS-51B was the first fully operational flight of the European-built Spacelab. The mission was called "Spacelab 3," even though it was the second Spacelab mission to fly—originally planned to fly third.

NASA's luck continued with the successful flight of the Discovery orbiter mission STS-51G launched June 17, 1985, at 7:33 A.M. (EDT). Carrying an international crew commanded by Daniel Brandenstein with John Creighton as pilot, the Discovery orbiter launched three communications satellites, conducted a series of "getaway special" experiments and launched and then retrieved Spartan 1, an astronomy spacecraft designed to be released from the Shuttle, perform observations for a few days and then be returned to the Shuttle before its trip back home. On board were mission specialists Shannon Lucid, Steven Nagel and John Fabian, along with payload specialists Patrick Baudry of France and Sultan Salman Al-Saud of Saudi Arabia. STS-51G returned to Earth and Edwards Air Force Base on June 24, 1985, at 6:11 A.M. (PDT) after spending a very successful seven days, one hour, 38 minutes and 53 seconds in space and performing 112 orbits.

Bad luck began to haunt NASA, though, on its next mission, STS-51F, the 19th Shuttle flight. Scheduled originally to launch July 12, 1985, the Challenger orbiter found its lift-off delayed for 17 days after a malfunctioning coolant valve on its number-two engine was discovered. More problems cropped up on its new launch date, July 29, 1985, when one of the Challenger's three engines shut down prematurely six minutes after launch as it was heading toward orbit. Normally the shuttle's three engines burn for about nine minutes to achieve orbit. But, with one down and the other two still burning, the crew had to jettison rocket fuel to lighten the load or risk a tricky emergency landing in Spain. Fortunately, Challenger managed its climb to orbit minus one burning engine. Commanded by Charles Fullerton with Roy Bridges as pilot, its crew and its Spacelab 2 payload performed a flawless mission before returning to Earth on August 6, 1985, at 12:45 P.M. (PDT).

The Shuttle orbiter Columbia comes in for a landing

A 1983 view of Challenger riding a Boeing 747 back to Kennedy Space Center in Florida from its California landing site. Johnson Space Center in Houston can be seen in the background

Spacelab—Home for Science in Space

Description: Spacelab is a special laboratory module designed and built in West Germany and financed by nine members of the European Space Agency (ESA)—Belgium, Denmark, France, Germany, Italy, the Netherlands, Spain, Switzerland, the United Kingdom—and one associate member, Australia. It cost $1 billion to build. Designed to fit in the Shuttle cargo bay, Spacelab modules are usually pressurized and astronauts can enter the laboratory from the pressurized crew quarters through an airlock and work in their shirtsleeves (even though the cargo bay containing the Spacelab is not pressurized and astronauts working there typically have to wear EVA suits).

History: STS-9 flew the first Spacelab mission in November 1983. The first non-astronauts to fly on the Shuttle were on board, West German physicist Ulf Merbold (who also was the first non-American to fly in the U.S. program) and biomedical engineer Byron Lichtenberg. It was their job

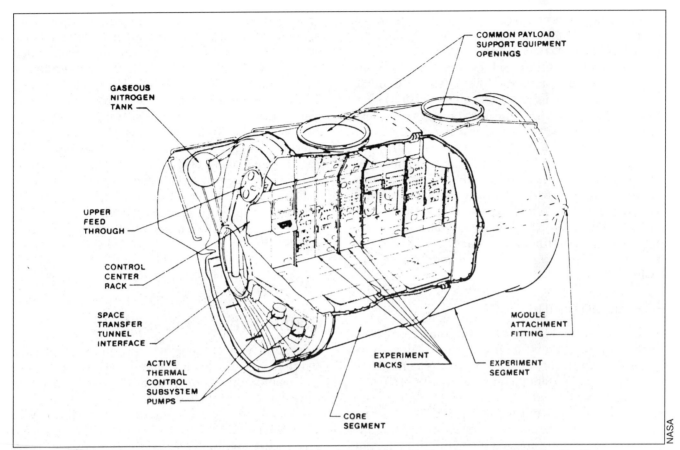

COMMON PAYLOAD
SUPPORT EQUIPMENT
OPENINGS

GASEOUS
NITROGEN
TANK

UPPER
FEED
THROUGH

CONTROL
CENTER
RACK

SPACE
TRANSFER
TUNNEL
INTERFACE

ACTIVE
THERMAL
CONTROL
SUBSYSTEM
PUMPS

MODULE
ATTACHMENT
FITTING

EXPERIMENT
RACKS

EXPERIMENT
SEGMENT

CORE
SEGMENT

NASA

A cutaway view of the European Space Agency's Spacelab module, designed to fly in the cargo bay of the Shuttle

to tend the 72 experiments flown on Spacelab 1—ranging from research in space medicine (especially the effects of weightlessness on the human body), astronomy, space physics, Earth physics and materials processing. From the ground, 200 scientists scattered around the world advised the crew on conducting the experiments, and, despite some annoying communications problems, the flight went well.

Spacelab 3, which was moved ahead in the schedule because of delays in the development of Spacelab 2, flew aboard STS-51B in May 1985. For the first time the Shuttle flew animals on board, and the opportunity enabled scientists to study the effects of microgravity on two rats and 24 monkeys. Of the 15 experiments, including crystal-growth and fluid mechanics research, 14 were successful. Materials scientists Lodewijk van den Berg and fluid physicist Taylor Wang, who helped design the experiments, did on-board research and tended the experiments.

Spacelab 2 finally got its chance in July 1985 on STS- 51F, a Challenger flight. It carried 13 experiments in seven scientific disciplines: solar physics, atmospheric physics, plasma physics, infrared astronomy, high energy astrophysics, technology research, and life sciences. This time Spacelab did not include pressurized modules, but provided unpressurized platforms for instruments, an "Igloo" that housed computers and electronic systems and a pointing system. The result was a unique orbiting observatory ready to study the Sun, stars and space.

NASA's next mission STS-51I with the Discovery orbiter also suffered a three-day delay when its original August 24, 1985, launch was scrubbed due to weather conditions. Rescheduled for August 25, an onboard computer failure delayed launch another two days until August 27, 1985. Once in flight after lift-off at 6:58 A.M. (EDT), the mission was a complete success. It included an in-orbit repair of the Leasat 3 satellite (launched originally by Discovery during its 16th mission in April 1985) and the deployment of three more communications satellites. With Joe Engle commanding and Richard Covey serving as pilot, STS-51I, NASA's 20th Shuttle flight, returned to Earth and Edwards on September 3, 1985, at 6:16 A.M. (PDT), after spending seven days, two hours, 18 minutes and 29 seconds in space and making 112 orbits.

Atlantis Joins the Fleet

Although the dream-weavers had seen the Shuttle system as a continuous flow of outgoing and incoming spacecraft capable of deploying payloads of materials or carrying passengers on an almost daily basis, the hard realities of economics and politics had caused endless cost-cutting and compromising. The promising Space Transportation System was limited to four over-worked orbiters and a system with little margin for slippage or error.

Joining the fleet for mission STS-51J, Atlantis, the last of the system's four orbiters, was put to work for the second secret military mission for the Department of Defense. Launched from Kennedy on October 3, 1985, at 11:15 A.M. (EDT) Atlantis spent four days, one hour and 45 minutes performing its secret duties and then returned to Earth without incident on October 7, 1985, at 10:00 A.M. (PDT).

The Challenger orbiter drew duty next for mission STS-61A. It was Challenger's ninth flight to orbit. Launched from Kennedy Space Center on October 30, 1985, at 12:00 noon (EST) STS-61A carried the first dedicated German Spacelab as its payload. Commanded by Henry Hartsfield and piloted by Steven Nagel, the mission also carried a record crew of eight, including mission specialists James Buchli, Guion Bluford and Bonnie Dunbar, as well as payload specialists Reinhard Furrer, Ernst Messerschmid and Wubbo Ockels. The first U.S. space mission to have its in-orbit research directed from another country (West Germany), STS-61A spent seven days, 44 minutes and 51 seconds in space, completing scores of scientific experiments and observations in its 111 orbits before returning to Earth on November 6, 1985, at 9:45 A.M. (PST).

The Atlantis orbiter took center stage again with mission STS-61B. Launched at night on November 26, 1985, at 7:29 P.M. (EST), the mission was commanded by Brewster Shaw with Bryan O'Connor serving as pilot, and a crew made up of mission specialists Mary Cleave, Sherwood Spring and Jerry Ross, plus payload specialists Rodolfo Neri (the first Mexican national to travel in space) and Charles Walker. STS-61B launched three communications satellites and promoted the feasibility of NASA's planned Space Station when Ross and Spring made two space walks demonstrating space construction techniques using snap-together parts to form a small space tower and pyramid.

After performing other scientific and engineering experiments Atlantis returned to Earth on December 3, 1985, at 1:33 P.M. (PST). It had made 109 earth orbits during its six-day, 21-hour, four-minute and 50-second stay in space.

Countdown to Tragedy

Television viewers saw nothing unusual as they watched what was for many becoming almost routine pictures of the Space Shuttle's lift-off. Some news commentators had even begun to speak of the program as being colorless and uneventful. The more observant, though, had watched the slow slippage in the program's schedule and felt the strain that ran throughout the space community.

The Shuttle program had been operating on a frayed shoestring. NASA struggled to live up to its highly publicized promises, but time and fate were uncooperative. There was little margin for error built into the program from the beginning and as the perhaps overly ambitious timetable began to unravel, the temptation was perhaps too strong to pull the shoestring tighter, and hope that it wouldn't break.

Things didn't get any better with STS-61C, the program's 24th mission.

The plan was for the recently refurbished Columbia orbiter to launch in late December 1985, deploy a communications satellite, run a dozen science experiments and take light-intensified photographs of Halley's Comet as it neared the Sun. On December 18, however, when NASA's overworked ground crew fell behind schedule, the mission experienced its first delay. Rescheduled to launch on December 19, STS-61C was less than 14 seconds to lift-off when the mission was halted because of troubles in one of the solid rocket boosters. Another delay on January 4, 1986, led to a January 6 reschedule when the mission was again halted, this time with 31 seconds left to go in its countdown. A faulty valve was corrected but the next planned lift-off,

ORBITER ACCOMMODATIONS
Seats, restraints, and mobility aids
Egress systems
Flight data file
Sighting aids
Photographic equipment
Window shades and filters
Stowage areas
Food systems and equipment
Sleeping accommodations
Crew hygiene systems and accommodations
Housekeeping equipment
Airlock

CREW EQUIPMENT
Survival equipment
Medical kits
Radiation instrumentation
Operational bioinstrumentation
Crew clothing
Space suit assembly

NASA

Crew quarters aboard the Shuttle

scheduled for January 7, 1986, was again rescheduled because of bad weather. On January 8, a day before the planned lift-off, another valve problem resulted in another reschedule to January 10 when weather problems again postponed the launch until January 12, 1986.

Finally, on January 12, 1986, at 6:55 A.M. (EST), seven delays and 25 days after its original scheduled launch, Columbia lifted off from Kennedy and climbed to orbit. It was a weary and unhappy crew, though, that returned six days later to make a pre-dawn landing on January 18, 1986, at Edwards Air

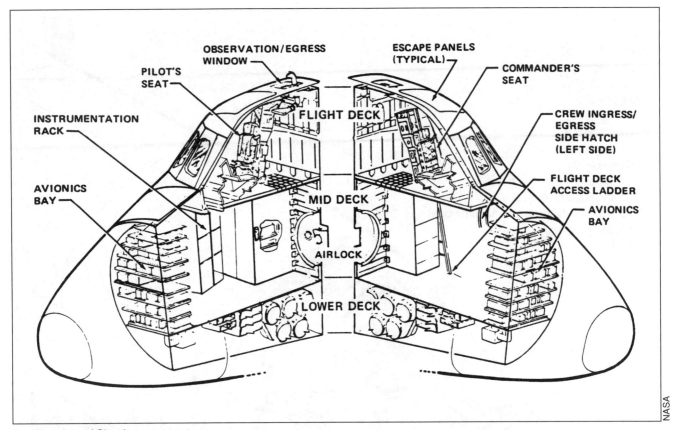

Another view of Shuttle crew quarters

Force Base. Although the communications satellite had been successfully deployed, run-down batteries in the camera prevented the hoped-for pictures of Halley's Comet from being taken.

Challenger: And the Challenge of the Future

STS-51L was to have been the 25th Shuttle mission. Its primary duties were to deploy the second Tracking and Data Relay Satellite (TDRS-B) and fly the Spartan Halley Experiment which would gather ultraviolet information from the coma (the cloud of dust and gas surrounding the nucleus) and tail of Halley's Comet. Its crew of seven included Francis Scobee as commander and Michael Smith as pilot, mission specialists Judith Resnik, Ellison Onizuka and Ronald McNair, plus two payload specialists, Gregory Jarvis and "teacher-in-space" Christa McAuliffe.

On January 28, 1986, shortly after 11:38 A.M. (EST), time and fate caught up with NASA. The frayed string broke and the American space program came to a sudden and shocking halt.

Seventy-three seconds after lift-off, in a sudden burst of flame and smoke trailing across the blue Florida sky, the Space Shuttle Challenger carrying its seven crew members exploded in the air and plunged to the waters below.

It was the worst tragedy in the history of the American space program. As the world mourned the deaths of the seven American astronauts, time stood still for the remainder of the Space Transportation System fleet, virtually the entire American space program.

The facts came out slowly in the lengthy investigation that followed.

Photographs taken shortly after ignition showed a small puff of black smoke coming from the lower end of the right rocket booster where synthetic rubber O-rings were supposed to seal the joint. Cold weather at the Cape had apparently caused the rings to become so brittle that they no longer could seal properly. The smoke disappeared briefly as Challenger cleared the launch tower, apparently when the burned rubber of the rings temporarily re-sealed the leak. As it continued its upward rise, Challenger then experienced a tremendous buffeting, due possibly to wind shear effects created by changes in wind direction, which caused the damaged seals to fracture again. This time the damage was apparently too great to re-seal. A jet

Pioneers who gave their lives: All seven members of the Challenger STS-51L crew died in an explosion just moments after lift-off. Left to right, seated: astronauts Michael Smith, Francis R. (Dick) Scobee and Ronald E. McNair. Standing: astronaut El-lison S. Onizuka, teacher-in-space Sharon Christa McAuliffe, payload specialist Gregory Jarvis and astronaut Judith A. Resnick

of flame burst from the joint and began to burn through the skin of the main fuel tank and the struts that attached the rockets to the tank. At 72 seconds after lift-off, the burned upper strut (or brace) severed, causing the pointed nose of the solid rocket booster to fall inward and crash into the giant external tank. In less than a second the liquid oxygen inside the ruptured tank began to pour out, igniting in a gigantic explosion of fire which literally ripped Challenger apart and plunged the spacecraft and its crew to the ocean below.

The mechanics of the event as outlined in the final report of the special commission set up to investigate the Challenger tragedy were cold and unemotional. But in the weeks of investigations and testimony that preceded the final report on the tragedy, another story, a more human and disturbing one, began to emerge.

The story was one of pressures and stresses, of weaknesses and failures, not in rockets, tanks and machinery but in human beings. Witnesses told of two years worth of disregarded worries and warnings about the fragile O-rings. Of alarm and concern about the cold weather and the brittle O-rings prior to the Challenger flight. About upper-level management pressures and impossible schedules pushing for a flight that morning that never should have been made. About narrowly avoided dangers in previous launches and overworked ground crews on the edge of exhaustion. About a dangerously flawed decision-making process within NASA, and broken and deliberately avoided lines of communication between

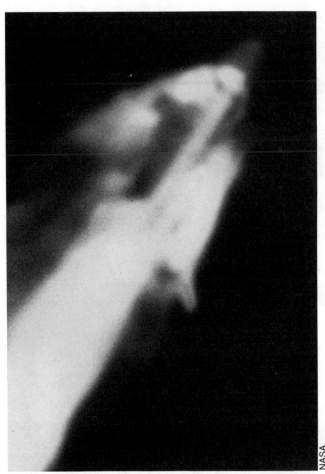

Challenger exploding 73 seconds after lift-off on January 28, 1986, killing all seven crew members and bringing the United States space program to a screeching halt for two and a half years of investigation and redesign

upper and middle levels of management, engineers and astronauts.

And Now Tomorrow?

NASA was badly shaken, both internally and in its public image. Now the space agency began a slow healing process after Challenger. Correcting the dangerous O-ring problems was only the first necessary step. In the months after Challenger, with America's space dreams on a long and painful hold, NASA struggled to regain its sense of direction, price and conscience.

It took two and a half years of careful rebuilding, with many questions asked and extensive testing and re-testing. But finally, on September 29, 1988, the orbiter Discovery thundered into orbit. The five astronauts on board were all veterans, with Frederick Hauck as commander, Richard Covey as pilot, and

George Nelson, David Hilmers and John Lounge as mission specialists. Their main job was to show that the Shuttle was back "up and running," and they did that beautifully, gliding in for a smooth and perfect landing on a clear California day at Edwards on October 3.

Two months later Atlantis followed with STS-27, a military mission lasting only four days. Launched on December 2 and returning on December 6, the crew of five launched a badly needed military surveillance satellite that had been grounded during the long wait after Challenger. Robert Gibson, Dale Gardner, Jerry Ross, Richard Mullane and William Shepherd were the crew members.

Beginning to breathe a sigh of relief, but still watchful for problems with safety, NASA faced an ambitious launch schedule for 1989. Several delays haunted the STS-28 lift-off, but Discovery finally got off the ground on March 13 with Navy Captain

Discovery rolls out to the launchpad shortly after midnight on July 4, 1988, in preparation for the first Shuttle launch in more than two and a half years, STS-26

Michael Coats in command. The other four astronauts in the crew were John Blaha, pilot, and mission specialists James Buchli, Robert Springer, and physician James Bagian. Six hours after a smooth lift-off the astronauts released a long-awaited TDRS tracking satellite that is vital to communication between the Shuttle and other low-orbiting spacecraft and the ground.

As the successes continued once again, Americans were living and working in space, free from the bondage of Earth. Seven courageous and visionary American astronauts who gave their lives had not done so in vain.

In the quest for space there is only upward. If humanity's future lies in the direction of the stars there can be no other way but through danger, with courage, grace and dreams.

7

THE SOVIET MIR SPACE STATION: TOWARD A PERMANENT HUMAN OUTPOST IN SPACE

It's like a huge seagull!
—Soyuz T-15 crew members as they viewed
Mir from orbit for the first time, March 1986

Salyut 7 had been ailing for some time when, less than a month after the United States suffered the loss of Challenger, at 12:28 A.M. Moscow time on February 20, 1986, the Soviets launched a new space station called "Mir," meaning "Peace." As the big SL-13 booster soared into the wintry sky at Tyuratam (also known as the Baikonur Cosmodrome, located in Kazakhstan in Central Asia) that day, it breathed ever new life into a space program that has been steadily gaining in technological expertise, experience in space and success. By 1987, the Soviet predominance would stand at a phenomenal 90 percent of all worldwide space operations with 91 space missions launched by the USSR during the previous year (an average of one every four days).

Much of the activity revolved around the new space station, which was opened less than a month after launch by Leonid Kizim and Vladimir Solovyev, who arrived on March 15. By May they had cruised in their Soyuz T-15 over to the co-orbiting Salyut station to dock at the rear port, later returning to Mir.

While some observers point out that the basic design of Mir closely emulates Salyut's, the Mir does have numerous operational improvements. Most sig-

nificant of all is the shortened transfer module (for entrance into the space station), which has a total of four docking ports around the sides, instead of the one port at the nose that the Salyut stations had. This difference alone means that several Cosmos laboratories or modules can be docked at the Mir at the same time—greatly multiplying the amount of experimental work that can be done. Because the Soviet Union has become more and more eager to welcome commercial use of their facilities, this feature of Mir may open up new possibilities for scientists who want to do micro-gravity work in astrophysics, Earth resources, materials processing and biology.

Mir's original two solar panels are also twice as large as Salyut's, providing at least nine to 10 kilowatts of power. Cosmonauts Yuri Romanenko and Alexander Laveikin (Soyuz TM-2), who took up residence on February 7, 1987, went out on two EVAs in June to install a third solar array—so the actual power level available on Mir must now actually be much higher.

Additionally, an antenna installed at the base of the station establishes communications contact between Mir and the Soviet data communications relay satel-

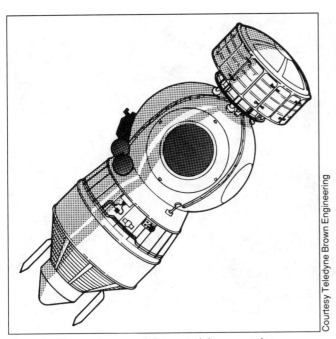

The new Photon spacecraft for materials processing— directly descended from the Vostok spacecraft flown by Yuri Gagarin in 1961

Courtesy Teledyne Brown Engineering

technicians can take care of routine "housekeeping" and other controls to free the cosmonauts for scientific experiments and other work, such as high-quality crystal growth, that can only be done in zero gravity.

Mir also boasts of improved living conditions, with experiments housed away from the living area in separate modules, and private phone booth-sized crew quarters for each cosmonaut. The new galley now "seats" six people, and Mir is generally more roomy than any previous Soviet space station although the pressurized volume (or living/working quarters) of the Mir is only about one-third of Skylab's.

On March 31, 1987, a 22.7-ton astrophysics module was launched from Tyuratam by a Proton booster, but had difficulty docking, and Romanenko and Laveikin performed an EVA to clear an obstruction out of the way. When docked, the new module increased the length of Mir by 50 percent.

Laveikin had hoped to set a new endurance record along with Romanenko, but a cardiovascular problem curtailed his stay in July 1987, when he was relieved by Soyuz TM-3 cosmonaut Alexander Alexandrov and returned home with Syrian cosmonaut Mohammed Faris and Soviet rookie Alexander Viktorenko. Romanenko stayed behind, and on October 1 (30 years after Yuri Gagarin's flight) he broke the record with 238 days in orbit. Romanenko remained until December for a total of 326 days in space.

lite—a major breakthrough in a program that has been dogged with ground-station communications problems due to the fact that the USSR does not have a widely spaced ground-based antenna system like the one used by the U.S.—or a satellite-based tracking system like the Tracking and Data Relay Satellite System (TDRSS) recently launched by the U.S to replace its ground system.

Eight computers on board Mir, instead of one, mean that more data can be processed at once, but they also create a level of output that necessitates increased communications with the ground as well. Ground

But Mir is by no means the only bright spot on the Soviet space horizon. The long-awaited launch in 1986 of the Energia heavy-lift vehicle, designed to lift very heavy loads into orbit, may be the key to developing even larger and more sophisticated Soviet space stations in the future. Its success may be a stronger

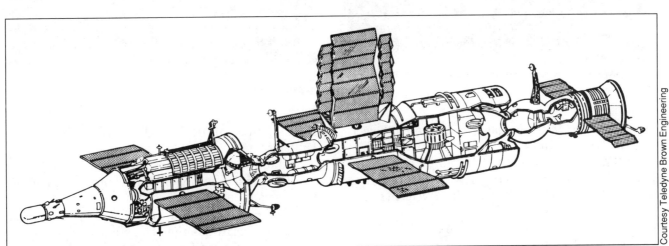

Mir space station docked with Cosmos 1443 (l) and Soyuz T (r)

Courtesy Teledyne Brown Engineering

Mir in flight

TASS/SOVFOTO

Kizim in working section of Mir during Soyuz T-15 mission in 1987

TASS/SOVFOTO

Romanenko and Laveikin on board Mir during Soyuz TM-2 mission, September 1987

TASS/SOVFOTO

indication that Mir is just an interim step to bigger, more ambitious things.

The Soviet Biosatellite Cosmos 1887, a satellite investigating the effects of zero gravity on primates, was launched on September 29, 1987, for a week in space, a follow-up to the work done on Spacelab aboard the U.S. Shuttle. Although the experiment suffered when a Rhesus monkey slipped out of his harness and began pulling on electrodes, it gained much cooperation from scientists throughout the world, including the United States.

Yet another piece falls in place as the Soviet space shuttle launches for the first time, providing another new transportation link to space.

The Soviets have built a powerful space program that's growing by quantum leaps, and any nation that wants to keep pace faces an enormous challenge in the years to come.

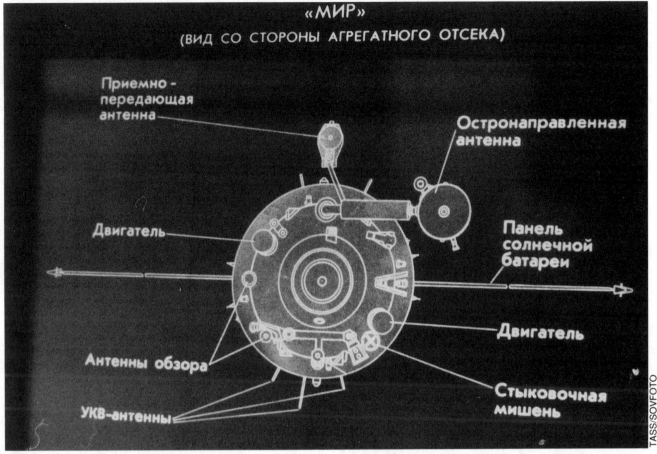

«МИР»
(ВИД СО СТОРОНЫ АГРЕГАТНОГО ОТСЕКА)

Приемно -
передающая
антенна

Остронаправленная
антенна

Двигатель

Панель
солнечной
батареи

Двигатель

Антенны обзора

Стыковочная
мишень

УКВ-антенны

TASS/SOVFOTO

Cross section of Mir

The Mir Space Station: An Overview

Launch: February 20, 1986 (Moscow time)
Overall length: 43 ft. (13.13 m)
Diameter (at largest point): 13.6 ft. (4.15 m)
Docking ports: Six, one aft, primarily for docking Progress cargo ferries, and five on the transfer compartment.
Compartments: Four. Three pressurized for human use—transfer compartment, working compartment and transfer adapter. One unpressurized service compartment for equipment and some supplies.
Weight at launch: 23 tons (20,900 kg)

The Mir space station, heralded as the first modular space station, consists of a core or central module the shape of two cylinders, one slightly larger than the other, connected end to end. At one end a sphere-shaped transfer compartment contains five ports—one at the forward end of the craft and the other four spaced around the circumference of the sphere. Crew and laboratory modules arriving at Mir dock at one of these ports and cosmonauts move into the space station (transferring from their spacecraft) through the transfer compartment, which is only 9.3 ft. (2.84 m) long.

Inside, the work compartment takes up the two main cylinders of the space station. The smaller of these two areas is 9.5 ft (2.9 m) in diameter (the wider being 13.6 ft. [4.15 m]) and the total length of the work compartment measures 25.16 ft. (7.67 m). Unlike the Salyut stations, Mir is designed to carry very little scientific equipment in this area, so that cosmonauts can use it primarily as living quarters and as the control area for the station's operations. Additional laboratory modules, such as the Kvant astrophysics laboratory, dock at the transit module to provide areas for conducting scientific experiments.

At the back of the station, a stubby cylinder 2.26 m long and 4.15 m in diameter contains the service compartment, which holds main engines, attitude control thrusters, antennas, lights, and docking targets.

Large, winglike arrays of solar cells provide power for Mir, the initial two panels carried outside Mir measuring 97.5 ft. (29.73 m) total span.

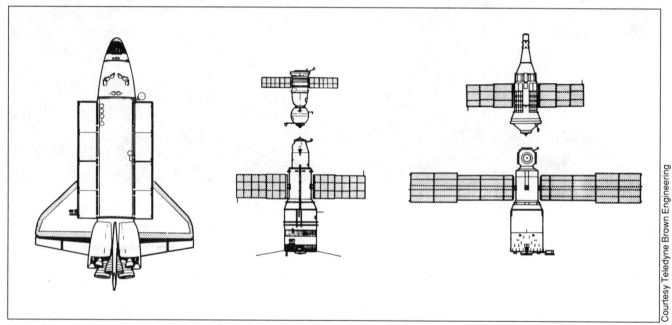

Mir and Cosmos module (right) compared to U. S. Shuttle (left) and Soyuz and Salyut 7

Courtesy Teledyne Brown Engineering

8

EXPLOITING THE NEW FRONTIER:
THE SPACE STAKES GO INTERNATIONAL

Although America's major political and exploratory competitor remains the Soviet Union, our major commercial competitors in space will be Western Europe and Japan.
—The Report of the National Commission on Space

As the United States and the Soviet Union continue their rivalry in space the rest of the world has also entered the great space stakes. With the growing realization of the commercial viability of space, particularly in the areas of communications and other applications satellites (such as scientific and earth resources), the space race has become not only scientific, political and military, but also economic as well. As space "exploration" has become space "exploitation," a growing number of nations have begun taking a closer look at the "apples in the space pie" and deciding that they might invest in a piece for themselves.

France Throws a Wide Space Net

In the aftermath of World War II, France, under the leadership of Charles de Gaulle, became one of the earliest European nations to enter the space age with de Gaulle's decision to push his nation into the forefront of the emerging technology sweepstakes. Deciding that a French presence in space would be an important element in achieving that ambition, the French government invested large amounts of money

in space research and the development of an efficient launch system. Just eight years after Sputnik 1, with the successful lift-off of its three-stage *Diamant* rocket on November 26, 1965, France joined the growing community of nations having launch capabilities.

Determined in its insistence that Europe should have its own launch capabilities and not be dependent on either the United States or the USSR, France developed the Diamant on its own while continuing to participate in the fledgling European Launcher Development Organization (ELDO) formed by Britain, France, Germany, Italy, Belgium and the Netherlands. ELDO had been set up in February 1964 as a seven-nation agency to develop and exploit the British "Blue Streak" rocket as a European launch system, but neither ELDO nor the "Blue Streak"-based launcher named Europa survived that rocket's many failures and the organization collapsed in April 1973.

When the highly successful European Space Agency (ESA) was organized in July 1973, France immediately became a driving force for the development of a new launch system and the prime developer of the

75

Ariane rocket (based on its Diamant experience) that would eventually make ESA a major space contender.

Continuing to build its own strong national space program, France founded CNES (the Centre National d'Études Spatiales) in 1962. France has remained the strongest participant and guiding force for the European Space Agency as well.

Casting its "space net" even wider, France has flown its own astronauts in both the USSR's Salyut program (June 1982) and the U.S. Space Shuttle program (June 1985). It has also begun negotiations with both the USSR and the United States to give the planned ESA project, the Hermes Space Shuttle, docking capability with the Soviet Mir space station and the planned U.S./International Space Station Freedom, which is to be built in the 1990s.

Great Britain Hesitates

If France was quick to recognize the importance of space, Great Britain has displayed a much more hesitant and conservative attitude. Despite the presence of one of the world's largest, strongest and most committed quasi-amateur space organizations, the British Interplanetary Society, whose history stretches back to the days of the earliest rocket experimenters, the British government itself displays a curiously unfocused and somewhat indifferent attitude to space exploration and exploitation.

Although a national space agency, the British National Space Center, was finally established in February 1985, financial cutbacks and political indecision have plagued the program almost from the start, leading to discontent and defections from the agency which continue to this day.

An indicator of the nation's indecisiveness toward space was the fact that, although Great Britain technically became the sixth nation (after Russia, America, France, Japan and China) to achieve independent launch capability with the successful lift-off of its Black Arrow launch craft in October 1971, the program, which put England's only completely national satellite (Prospero) into orbit, was cancelled almost immediately after the flight.

Despite the faltering nature of its own independent program, Great Britain continues to be a prime participant in the European Space Agency as well as a major manufacturer of space satellites launched by other countries. Great Britain has a long history of cooperation with the American space program that stretches back to the launching of the ARIEL-1 scientific satellite in the spring of 1962, and the British also built the hardware and supplied ground support for the successful International Ultraviolet Explorer

Artist's conception of the Japanese H-2 rocket

satellite, with its highly sensitive ultraviolet telescope for observing astronomical emissions beyond the high range of the visible spectrum, which was launched in 1979. They have participated with West Germany, the Netherlands and the United States in developing the Gamma Ray Observatory (a satellite designed to make a comprehensive astronomical study of gamma ray emissions in the universe) and was a prime contractor in Europe's Giotto spacecraft mission, which intercepted and studied Halley's Comet in March 1986.

Japan Takes a Big Bite

Under the guidance of two separate, dedicated space agencies, Japan has become the world's third greatest space nation. The National Space Development Agency (NASDA) was founded in October 1969 and is responsible for the development and launch of communications and applications satellites. The Institute of Space and Astronautical Science (ISAS), reorganized in 1981, is responsible for scientific satellites and launch vehicles. Japan has surpassed even the European Space Agency in numbers of successful launches and plans to challenge ESA's commercial launch activities with its successful "H" series of rockets, especially the presently "in-development"

H-2 which could be a stiff competitor for ESA's Ariane launcher.

Independently launching its first national satellite OSUMI in February 1970, Japan currently has a space budget amounting to over $1 billion annually—and unlike the U.S. space budget, Japan's appears to be well secured into the future. A strong participant in the U.S./International Space Station program, its contributions already show great promise.

With its strong technological base, especially in computer and artificial intelligence technologies, Japan is in an ideal position for continuing to gain important new information from international technology transfers and shows every indication of becoming an even greater force in both space exploration and exploitation.

China Plays Its Hidden Hand

In 1986, shortly after the explosion of the American shuttle Challenger, China distributed brochures and "user's manuals" to the world's leading space players offering its advanced launch capabilities to any and all takers. Although it had made similar overtures in offering commercial launches two years previously, the temporary shutdown of the American space program after the Challenger disaster decidedly moved China into the high-stakes game of space exploitation.

Promoting its new Long March rocket system, by the end of 1986 China began attracting its first international commercial customers and, despite ideological problems with many of the "free world" nations, has built a slowly increasing commercial space business.

Ironically, much of China's space-launch capability is believed to have been made possible when the United States expelled the highly trained space engineer and scientist Tsien Wei-chang during America's notorious "McCarthy Era" of the late 1940s and early '50s, when Senator Joseph McCarthy led a rampage of anti-communist investigations and purges. Before his enforced return to mainland China, Tsien had received his Ph.D. from the University of Toronto and had worked as a top-grade research engineer at the Jet Propulsion Laboratory at Pasadena, California, in the United States.

With its launch of China 1 in April 1970, China became the fifth country in the world to achieve independent launch capability (not as part of a group or consortium). The satellite carried a small transmitter which broadcast a Chinese song "The East Is Red" paying homage to Chairman Mao, as it passed over the other nations of the world.

As a part of its continuing growth in the international space stakes, China announced in 1988 that it will join Brazil in a joint venture to produce and launch a series of Earth resources satellites intended to compete with the U.S. Landsat series.

A Cooperative Effort: The European Space Agency Rides Ariane to Success

The Chinese were not the only ones to try to quickly capitalize on the American Challenger disaster and the subsequent halt of the American space program. There were still satellites to be launched—communications satellites, science satellites, Earth resource and navigation satellites—and few launchers to launch them. By early 1988 even the Soviet Union was making a bid to attract commercial satellite launching. The European Space Agency, which a few years before was struggling to live down some costly launch failures and gain a reputation of reliability, soon found itself swamped with business.

Formed in July 1973 by 11 nations after the collapse of the ill-fated European Launch Development Organization, the European Space Agency looked as if it might follow ELDO's rocky road to failure. ELDO had depended on the Europa launcher to get off the ground, and the organization crumbled with the rocket's repeated failures. ESA similarly pinned its hopes on the French promoted Ariane launcher based on France's Diamant rocket.

For a while it looked as if ESA's chance of success was "iffy" to say the least. After a series of delays and expensive launch failures, the first of the planned Ariane series, Ariane 1, was declared operational in January 1982. However the program continued to meet with a spotty succession of delays, successes and expensive failures. One of its most embarrassing was a highly publicized launch attempt in September 1985 which had to be destroyed by the safety officer in view of French President François Mitterrand who was observing from the launch control tower.

Still Ariane accounted enough success to keep it and ESA in the space stakes. The Arianespace Consortium, which had taken over the production and marketing of operational Ariane rockets for ESA after the ninth launch, found itself besieged with launch requests after the United States announced a halt on commercial launches following the loss of Challenger. This despite the fact that in May 1986 Ariane V18 (and its costly satellite payload) also had to be destroyed by the launch safety officer, putting the entire Ariane program out of the game for over a year of study and redesign.

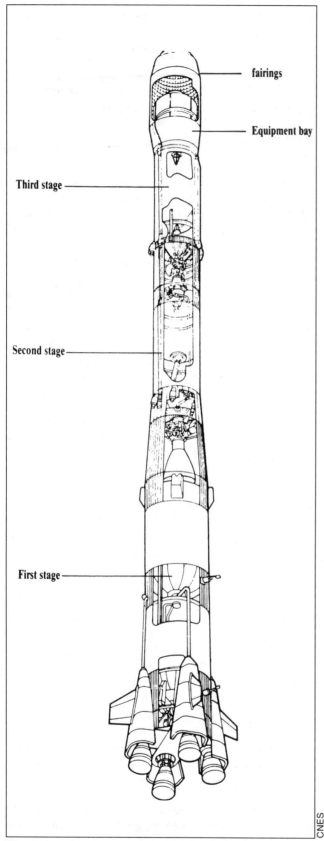

fairings

Equipment bay

Third stage

Second stage

First stage

CNES

The three-stage Ariane rocket design

Today, with a staff totalling over 1,300, including more than 600 skilled engineers and 120 scientists, ESA is riding high in the commercial space stakes although it feels the pressure of Japan closing in from behind. In an effort to keep ahead, it successfully launched the first of its new series of Ariane launchers, Ariane 4, in June 1988. Capable of mixing and matching its strap-on boosters to fit the requirements of individual payloads, Ariane 4 is planned to be the newest and most versatile workhorse of the Ariane fleet. Arianespace and ESA have also announced the go-ahead for Ariane 5 to be operational in the 1990s. If everything goes according to schedule, Ariane 5 will be capable of placing ESA's planned Hermes space shuttle in orbit, giving ESA and its member nations space shuttle capabilities similar to those now offered only by the United States and the USSR.

In addition to its commercial and special applications satellite launches, the European Space Agency has an active scientific program that requires mandatory participation on the part of its member nations. Among its many projects have been the International Sun-Earth Explorer Program (ISEE) in cooperation with NASA (1977), the International Ultraviolet Explorer (IUE), a joint NASA, ESA and United Kingdom project launched by NASA in 1978, and the strikingly successful Giotto flyby of Halley's Comet on March 13, 1986.

Like the French Space Program, ESA also bridges the political gap, flying missions with both the United States and the USSR. Typical of ESA/USSR cooperation was the April 13, 1987, docking at the Soviet space station Mir of ESA's Roentgen Laboratory, a small orbiting observatory built by ESA and launched by Russia.

ESA was also a billion-dollar contributor (somewhat unhappily so) to the sparsely used U.S./International Spacelab program flown on the Space Shuttle. The program has been plagued by delays and disappointments. The frustration with their Spacelab experience has also made ESA and its member nations uneasy and a little bit wary about their possible participation in the planned U.S./International Space Station Freedom, which has received lukewarm support in the U.S.

Upcoming science missions for ESA include the Ulysses project (a planned ESA/NASA solar Polar mission) and ISO (an Infrared Space Observatory to be launched by Ariane 4 sometime in the middle 1990s). ESA has also contributed almost 15 percent of the cost of the U.S. Hubble Space Telescope in return for nearly 15 percent observing time for European astronomers.

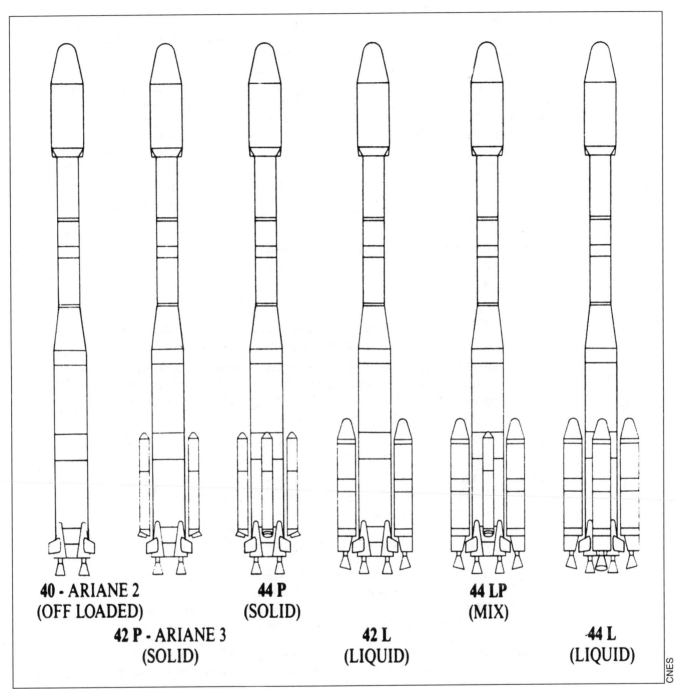

40 - ARIANE 2
(OFF LOADED)

44 P
(SOLID)

42 P - ARIANE 3
(SOLID)

42 L
(LIQUID)

44 LP
(MIX)

44 L
(LIQUID)

CNES

The different Ariane 4 launcher configurations

Taking Our Planet's Vital Signs:
International Observation of Earth from Space

When Apollo 8 astronauts Frank Borman, James Lovell and William Anders snapped the first pictures of our swirling blue and green planet from space, they changed forever our view of Earth. Suddenly we began to realize how small and fragile our little planet really is, and we began to gain a new sense of stewardship about our environment and the way we change it. Today, with loss of ozone in the upper stratosphere, the warming effects of carbon dioxide and other gases in our atmosphere, the loss of large expanses of tropical rain forests, and the disappearance of entire species of plant and animal life, more and more scientists and governments the world over are beginning to see problems like these as symptoms of an ailing planet. International concern about the state of the Earth, how the Earth's systems interrelate and what effects we're having on it has resulted in coordination among many national space programs and international consortia to gather much-needed data by making measurements from space. Below are listed just some of the national and international Earth observation projects from space now in effect or planned for the near future.

National Programs:

Brazil   **Remote Sensing Satellite**   Land applications satellite to launch in 1990/1991

Canada   **Radarsat**   Ice- and land-monitoring satellite planned to use radar instrumentation and to launch in 1994

China   **Earth Resources Satellite**   Land applications satellite using a multispectral scanner from sun-synchronous orbit. Due to launch in 1989

France   **SPOT-1**   Système Probatoire d'Observation de la Terre-1: France's first land use and Earth resources observation satellite, using a high-resolution imager. SPOT-2 will launch by the end of 1989, and two others are planned

India   **IRS-1**   Indian Remote Sensing Satellite: India's Earth resources satellite, part of a projected series

Japan   **ADEOS**   A planned Japanese sun-synchronous satellite due for launch in 1994, to make environmental observations of Earth and ocean resources in the visible and near infrared regions of the spectrum
GMS   Geostationary Meteorology Satellite: Japan's weather observation satellite, managed by NASDA
JERS-1   Japanese Earth Remote Sensing Satellite-1: Japan's Earth resources satellite, due for launch in 1991
MOS-1   Marine Observatory Satellite-1: Japan's satellite observatory studying the state of the sea surface and atmosphere

United States   **DMSP**   Defense Meteorological Satellite Program: U.S. Air Force weather observations with several instruments; measurements include microwave (very high frequency radio waves) and infrared radiation
ERBS   Earth Radiation Budget Satellite: A NASA/NOAA satellite currently observing the Earth radiation budget—contrasts data with polar-orbiting satellites NOAA-9 and NOAA-10
GOES   Geostationary Operational Environmental Satellite System: U.S. NOAA satellites making environmental observations, especially of wind speeds (more planned for the 1990s). Uses multiple instruments
GEOSAT   Geodesy Satellite: U.S. Navy satellite measuring shape of the Earth, as well as oceanic and atmospheric properties, using a radar altimeter

80

GPS Global Positioning System: A multi-agency U.S. system eventually to include a system of 21 satellites using transmitters to determine location precisely for studying the shape, size, gravity and magnetism of the Earth, and deformations in the Earth's crust

GREM Geopotential Research Explorer Mission: U.S. project currently on hold; would make highly accurate studies of the Earth's gravitational and magnetic fields

LANDSAT Land remote-sensing satellite: Satellite series that has already collected much data on vegetation, crop and land use inventory. Landsat missions 6 and 7 are planned for the 1990s

Nimbus-7 NASA satellite monitoring atmospheric pollutants, ocean chlorophyll concentrations, weather and climate

POES Polar-orbiting Operational Environmental Satellites: A series of U.S. NOAA environmental observation satellites (also known as the NOAA series of satellites) with follow-on missions planned for the 1990s. Uses multiple instruments

TREM Tropical Rainfall Explorer Mission: A U.S. planned program to start in 1991

UARS Upper Atmosphere Research Satellite: A NASA satellite planned for launch in 1991 to study stratospheric chemistry (including ozone depletion), dynamics and energy balance. Uses multiple instruments

Space Shuttle During U.S. Shuttle flights astronauts run Earth systems studies using various instruments including:

ATMOS Atmospheric Trace Molecules Observed by Spectroscopy: To study atmospheric chemical composition

ACR Active Cavity Radiometer: To measure solar energy output

SUSIM Solar Ultraviolet Spectral Irradiance Monitor: To make ultraviolet solar observations

Hand-held Camera Exploratory observations of meteorological oceanographic, biological and geological processes

Large Format Camera Detailed studies of land-surface features

LIDAR Light Detection and Ranging instrument: To measure surface topography, atmospheric properties

USSR **METEOR** Soviet Meteorological Satellite: USSR's meteorological observatory, measuring sea-surface temperature, sea ice, snow cover and vegetable conditions

International:

ESA **ERS-1** Earth Remote Sensing Satellite-1: An ESA satellite launched into sun-synchronous orbit in March 1989; carries multiple instruments to image oceans, ice fields and land areas
METEOSAT ESA's currently operational geosynchronous weather observation satellite

U.S./ESA/Japan **EOS** Earth Observing System: A complex system of instruments and platforms for observing changes over a decade or more in all the Earth's systems—ocean, land, biological species and atmosphere—simultaneously. To consist of four polar-orbiting platforms (two U.S. platforms (the first to launch in 1996), one ESA platform and one from Japan), several geosynchronous platforms and instrumentation aboard space station Freedom. They will coordinate with data gathered at the same time on the ground. To be in place by the year 2000, when the multi-instrument system will be processing 1 terabyte (1 trillion bytes) of data each day

U.S./France   **TOPEX/POSEIDON**   Ocean Topography Experiment: A joint U.S. and French project to study ocean circulation; to launch in 1991

U.S./Italy   **LAGEOS-1**   Laser Geodynamics Satellite: NASA satellite measuring geodynamics and gravity field using laser reflectors. Another being built by Italy due to launch in the 1990s

Some Smaller Bites: An Informal Roll Call of Space Interests

Although India became the seventh nation to achieve independent orbital capabilities in 1980 with its launch of Rohini 1B, an experimental satellite, the Indian involvement in space rests primarily on its continued cooperative participation with the U.S., USSR and ESA, having launched satellites through all three major space programs. India's present space plans call for complete orbital independence by the late 1990s but financial difficulties may delay its emergence as a major space power beyond that target date.

While the biggest pieces of the space pie have gone to the larger nations with independent launch facilities, many other nations have shared in the benefits and profits of space through cooperation and participation with ESA, or the larger American, Soviet, Japanese, Chinese or French space programs.

Among the more active of these, Canada has produced over a dozen scientific and communications satellites launched by NASA, and has participated in the U.S. Space Shuttle program by building the remote manipulator system used by several astronauts during Space Shuttle flights. The Canadians have also flown an astronaut as a payload specialist on NASA's Space Shuttle and are major contributors to the U.S./International Space Station.

The Federal Republic of Germany, with its increasingly active space program, has a long history of cooperation and participation in NASA programs, including acting as scientific ground controller in the

Spacelab D1 mission in November 1985, which was the first manned mission to be managed in flight by a country other than the United States or USSR. As a part of their country's active European Space Agency involvement, the West Germans have contributed nearly 30 percent to the preliminary development program of the proposed French-initiated Hermes spaceplane while also proposing a spaceplane project (called Sanger) of their own.

Brazil quickly took a strong interest in space communications and applications satellites. In the early 1960s it established two specific space agencies, INPE, the Institute for Space Research, responsible for handling civilian space research, and the IAE, the Institute for Space Activities, responsible for rocket development. The launch of Brasilsat 2 in May 1986 put Brazil in the telecommunications business when it was able to offer television channels for lease or sale to its neighboring countries. In 1988 Brazil announced a joint venture with China to produce and launch a series of earth resources satellites to compete with those of the U.S. Landsat series. In the past, Brazil has operated a Landsat ground station and has made extensive use of Landsat information for agriculture and other land management applications.

In May 1988, the Italian government gave final approval to the formal creation of an Italian space agency to oversee domestic satellite programs and to coordinate projects with the European Space Agency. The official organization, allocated over $1 billion to operate in its first couple of years, has been named Agenzia Spaziale Italiana (ASI), and will actively participate in the ESA Ariane 5 and Hermes space

CNES

Artist's conception of Ariane 4 launch

Courtesy Government of India Department of Space

India's Aryabhata satellite

shuttle projects. The formation of Italy's official space agency points out Italy's long interest and involvement in worldwide space activities, including its contributions to ESA and its major involvement in the Spacelab.

Among the other nations of the world participating in major ways to the worldwide space activities, Belgium has contributed experiments to Spacelab and made contributions to the Hubble Space Telescope program. Denmark was a major investigator in the High Energy Astronomical Observatory (HEAO 3) project launched in September 1979. The Netherlands has been involved in both ESA and NASA projects, including the infrared astronomical satellite (IRAS) launched by NASA in 1983. Norway and Switzerland have also contributed to the Spacelab project and Israel and Spain have both been involved in NOAA's Crustal Dynamics project which uses satellites and mobile ground stations to study movements of the Earth's surface.

Intercosmos: The Soviet Space Cooperative

The highly successful Intercosmos is the Soviet Union's program to involve Soviet bloc nations in the Soviet space program. Begun in 1967, the first Intercosmos participants were Bulgaria, Czechoslovakia, Cuba, the German Democratic Republic, Hungary, Mongolia, Poland and Romania. Cuba later became an active participant and non-Soviet bloc nations Spain and France have also been involved. With particular attention to the areas of space physics, communications, meteorology, biology and medicine, Intercosmos employs both manned and unmanned spacecraft, and representatives from nations participating in the program throughout have flown as cosmonauts to the Soviet Salyut and Mir space stations.

Intelsat: The First International Communications Satellite Giant

For Intelsat (the International Telecommunications Satellite Consortium), the business of space is business.

Founded in 1964 as a 14-country international consortium, Intelsat today has over one hundred member nations. Since its first successful satellite Intelsat 1 (better known as Early Bird) was launched in April

1965, it has launched and operated dozens of high-quality commercial telecommunications satellites.

The most successful commercial space enterprise in the world, Intelsat has put space to work on a large and profitable scale. Though beginning to face competition from new and smaller commercial operations, Intelsat holds a dominant position in telecommunications satellite technology.

Since recognizing the commercial benefits of space in the early 1960s it has quite literally changed the way the world sees itself. Intelsat satellites have beamed sounds and images across nations and oceans, giving almost instant access to history as it is being made. Its satellites have flashed the Olympic games before our eyes, as well as the horror of terrorism. They have made possible such events as the 1985 "Live Aid" concert for Ethiopian famine relief and they daily and efficiently handle over two-thirds of all the world's international telephone calls.

Across the Boundaries: The Space Above the Lines

The success of ESA, Intelsat, and other internationally cooperative space interests have shown the benefits of such international cooperation. The successful exploration and exploitation of space can be of immense profit to all nations on Earth, in knowledge, education, national prestige and hard dollars. It can also be enormously expensive. Cooperative ventures can both ease the financial load,

and share the profit. Perhaps more importantly such ventures may also extend the vision and possibilities of what the world of tomorrow may become.

The space giants of today, however, still remain the United States and the Soviet Union. The steps the giants will take are uncertain. Will space become truly a new frontier, a place where mankind lives and works or a final battlefield?

The USSR continues to push its space plans forward, not just toward permanent habitation of space but toward the continued exploration of the planets of the solar system. Unfortunately, its military space program also becomes more vigorous daily. There is little doubt, barring massive changes in the political direction of the nation, that the Soviets are in space to stay.

As the United States today ponders its space future, caught between administrative indecision, cost cutting and the increasing military pressures on its civilian space program, it is beginning to show signs of losing its prestigious place as a leader in space exploration and exploitation. Whether America will continue to move out into the "new frontier" to live and work, is a story only the future can tell.

Despite the uncertain political climates, the stops and starts, the hesitancies and the missed opportunities of any individual nation, the space age is here to stay. Space today is international. The doors have been opened; we have crossed the threshold, and there can be no turning back. The leaders of tomorrow will be those nations that choose today to invest in space and its immense potential.

EPILOGUE:
THE HUMAN FUTURE IN SPACE

Many children and young people today expect and anticipate that they will live and work in space. We can be confident that they will be fully as capable of taming that new frontier environment as were our ancestors who built America.
—Report of the National Commission on Space

You have heard it said that if God had wanted Man to fly, he would have given us wings. Well, today we can say that if God had wanted our species to produce a space-faring nation, he would have given us...the Moon!
—Krafft Ehriche

Tomorrow is not far away. The fruit of the future lies in the seeds of the present. Today hundreds of satellites and a small handful of humans inhabit space. Tomorrow the space outposts will expand and grow.

As the great Soviet rocket pioneer Konstantin Tsiolkovsky wrote, "Earth is the cradle of mankind but mankind cannot stay in the cradle forever." Will the child emerge healthy from the cradle, to grow to adulthood wise and strong? Will it reach out in curiosity toward new experiences, chart new paths, prepare new futures for itself and others? The fruit of the future lies in the seeds of the present. (Today, given humanity's will, we plant and nourish carefully.) Tomorrow, we will reap the harvest of advances, knowledge and new opportunities.

Permanent Space Stations

The U.S./International Space Station

Unable to maintain its orbit, America's first space station plunged back to Earth in 1979. One small piece

had, in an absurd defiance of the odds, struck and killed a solitary wild jackrabbit in the wilds of the Australian outback. The bird with the broken wing met its end in the Australian dust. Some said it should never have flown. Others said it should never have been left to die.

Though makeshift and improvised from the beginning, Skylab had been a home in space for a small handful of American astronauts, a man-made place where American astronauts could live and work. Skylab had been a way station in the vast darkness, an outpost and a symbol that said—"we are here, and we hope to stay here." Somehow. For many, its demise seemed to mark a turning point for the U.S. space program. After the death of Skylab, many observers thought that NASA became too enraptured with means over ends, with machinery rather than uses to which that machinery could be put. The Space Shuttle seemed to say it all. It was a Space Transportation System. It delivered satellites, picked them up and refurbished them in space. It carried scientific experi-

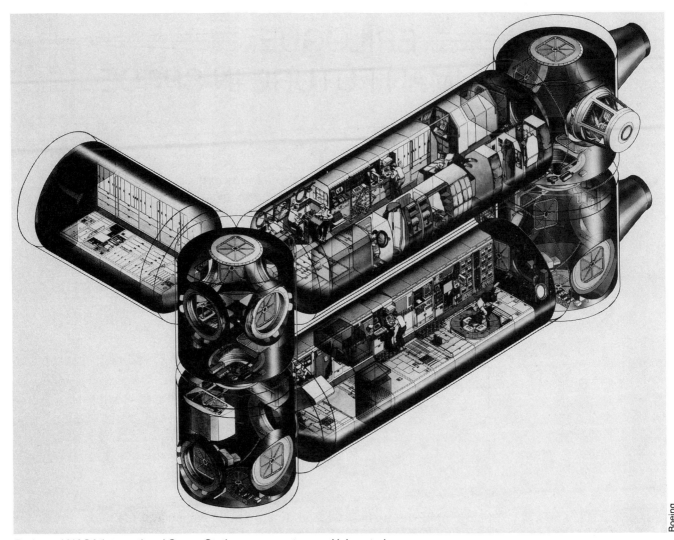

Boeing

Projected NASA/International Space Station crew quarters and laboratories

ments and scientists—even a special laboratory called Spacelab—for several days at a time. But Americans had no permanent place in space. No Salyut or Mir.

The U.S./International Space Station Freedom will change that, and give America a permanent place in space.

The U.S./International Space Station, however, will bear little resemblance to the gigantic, slowly rotating wonders of science-fiction books and movies. With its basic trusswork construction of struts, girders and attached modules, it will look more like the "Tinker Toy" or "Erector Set" project of a child's imagination than the high tech fantasy of a Hollywood motion picture director.

The design of the station will allow for simple, basic construction techniques and future expansion. The building parts of the station will be placed in the cargo bay of the Space Shuttle orbiter where they will be carried to low-Earth orbit to be unloaded in space by working astronauts and assembled in place over a period of months.

Operating in low-Earth orbit, about 250 miles high, the station will initially be composed of two primary modules, the living quarters, or habitat, module for the crew, and a laboratory module for scientific work. Later other modules will be attached as needed or desired, and so-called "free floating" platforms sharing the station's orbit and close by but unattached will also eventually become part of the station complex.

Inside the habitat and laboratory modules the astronauts will live more like submarine crewmen or remote-outpost scientists than luxury-liner passengers. Their living and working quarters will be small, cramped and maximized for the greatest use

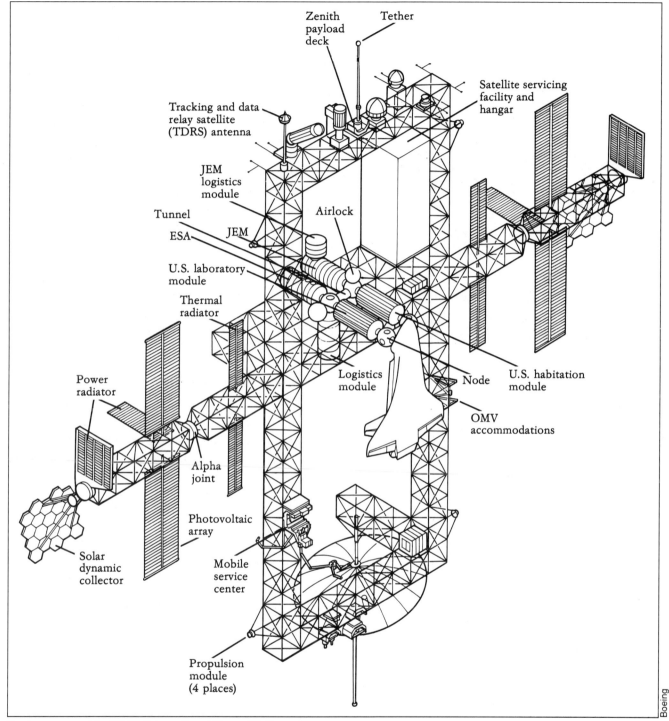

Design proposal for NASA/International Space Station

Boeing

and flexibility. Although each member of the crew will have a small closet-sized personal quarter to sleep, unwind or even work in (via a computer interface), privacy in the station will be rare and a spirit of mutual cooperation and self-discipline will be highly valued personal traits.

Astronauts inside the station will not experience the comfortable artificial gravity that allows science-fiction heroes like Captain Kirk and Science Officer Spock to move about their spacecraft with normal ease. Like their counterparts in the Soviet Salyut and Mir space stations, the U.S. astronauts will float and

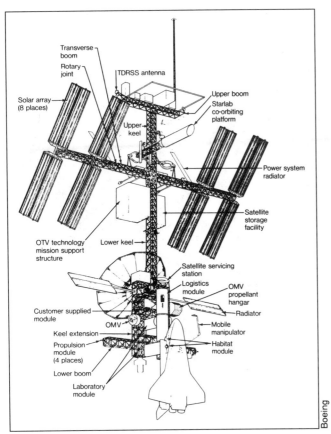

Design proposal for NASA/International Space Station

bob about in a zero-gravity environment, a situation that might ruffle the dignity of the ultra-cool, dignified and always-in-control Mr. Spock.

Although of a simple "bare bones" construction, the U.S./International Space Station will truly keep Americans "living and working" in space on a daily basis. Unlike the U.S. Skylab this station will be specifically built and outfitted for continuous and unbroken occupation by rotating crews of American and international astronauts.

Once in operation, the space station will give the United States and its international partners a permanent base to experiment with new space-based technologies, materials and drug processing and a way to study the mysteries of space and of Earth. The later addition of an "artificial gravity research facility" to the station complex will allow scientists to study the problems of creating an artificial gravity environment for extended space missions.

Eventually, too, the station may become a staging point for exploration of deeper space and our planetary neighbors, including a return trip to the Moon by American and international astronauts and our first visit to the planet Mars, "in person."

Future Soviet Space Stations

The Soviet do not plan to stop with their Mir space station design, either. They have discussed plans for a greatly expanded space station on the same scale as the one planned by the United States and the international community. Much more attracted to commercial interests in space in recent years, the Soviets will probably offer facilities for materials processing and manufacturing, as well as docking berths for scientific laboratories.

The Soviets have also described plans for Oblakos, "Swarms," separate laboratory platforms orbiting near each other, but not attached. These would be serviced by cosmonauts traveling by spacecraft from one to the other to take care of tending experiments or other housekeeping duties on board. But they would have no living quarters and probably no life-support systems. This arrangement might make a large space station complex unnecessary, or it might be built in addition to a more centralized facility.

The Big Vision: Colonies in Space?

Although controversial in the space community, particularly with many scientists who believe that "unmanned" space missions are the simplest and most economical way to explore the scientific mysteries of space, the U.S/International Space Station is for many others only the first step in a much larger and more ambitious vision. To really "live and work in space," these visionaries believe, humankind must create the large-scale, city-sized, space colonies of science-fiction dreams. Such scientists as American physicist Gerard K. O'Neill have explored plans for such "colonies," which include entire neighborhoods of individual housing, factories and farms, hospitals and office buildings, gigantic recreation centers and even artificial replications of forests and waterfalls.

Such large-scale space colonies, though, are visions still many years away from being realized. The hurdles standing between such dreams and their reality remain numerous. Scientists and engineers have much to learn yet about the problems inherent in fulfilling such dreams, problems such as "artificial gravity," ecological and environmental systems suitable for such large, long-duration habitation in space, and perhaps the biggest problem of all, convincing hesitant governments and industries that such dreams are economically and political desirable and feasible.

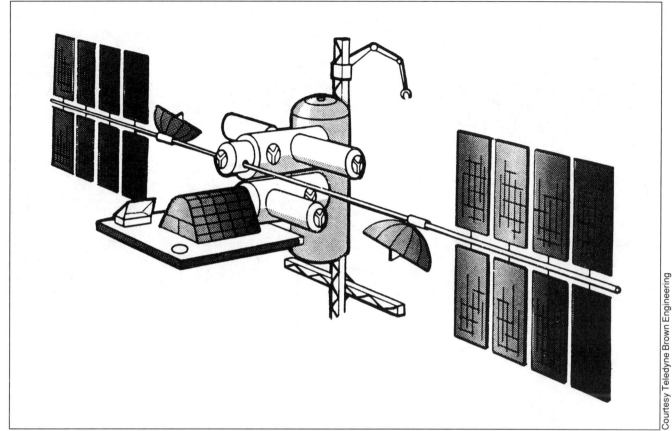

Courtesy Teledyne Brown Engineering

Concept (not intended to be accurate) for Soviet space station of the future, including possibility of a hangar for reusable spacecraft

Getting There from Here: Space Planes and Beyond

In the beginning it was the wagon trains that carried Americans to the Old West. Later, as the small settlements became established towns, a simpler, faster and more direct link was needed to keep the thriving ranches and businesses alive. The link that permanently connected the West to the rest of America was the railroad, with regularly scheduled runs, efficiency, relative low cost, and dependability.

Today, there are only two ways to send a human to space: expendable launch rockets, such as those used by several countries including the United States and the USSR (that's how a TM-Soyuz gets to Mir) or a space shuttle (like the American Columbia or the Soviet Buran). But, if near space is to become a thriving new frontier for humanity, then a more regular and efficient link between Earth and orbit is necessary. Toward that end many of the Earth's pioneering space nations are now studying advanced transportation systems to forge a better link to that new frontier.

The National Aerospace Plane

The United States is today studying the feasibility of a new generation of space transportation system—the National Aerospace Plane. A true one-stage Earth-to-orbit spacecraft, NASP will push current space technology to its limits, but if it succeeds it will bring the dreams of space pioneers one step closer to reality, opening up efficient and easy access to near space.

The key to NASP lies in its new propulsion technology—the "scramjet engine." Unlike ordinary rockets, which must carry the oxygen they need for combustion in gigantic tanks, adding to their weight, scramjets "breathe" oxygen from the air on their ascent to orbit, reducing their weight and allowing them to carry larger payloads with greater fuel economy.

Although designed to take off and land like a jet airplane from a conventional runway, NASP is no ordinary jet. In today's jet aircraft, air flowing through the aircraft's engines is compressed and heated by burning fuel. Then, by escaping and expanding through the exhaust nozzle, the air flow provides the thrust that pushes the aircraft.

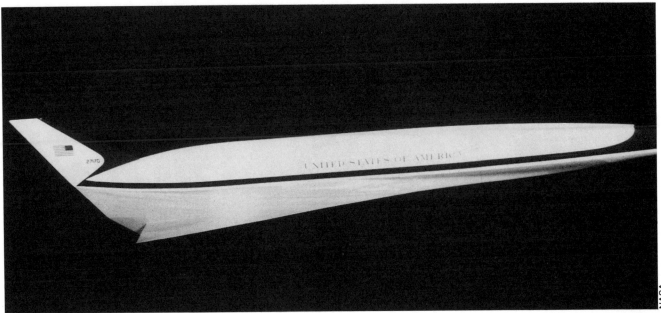

The U.S. National Aerospace Plane (X-30) of the future

The Japanese are designing an unmanned winged space vehicle, H II Orbiting Plane (HOPE)

The "scramjet" is a refined version of "ramjet" technology. After achieving speeds of Mach 1 (the speed of sound, 760 mph at 32 degrees F at sea level) with conventional technology, the ramjet relies on the aircraft's forward motion to literally ram air into its front opening, gulping it in at high speed. Once inside the aircraft, the incoming air slows down to subsonic speeds to mix with the fuel and ignite and escape through the rear nozzle. One of the major problems with ramjets, though, is that once reaching speeds near Mach 6 (six times the speed of sound) the temperature inside the aircraft's combustion chamber becomes so high that much of the fuel is expelled through the exhaust before it can become completely burned, cutting down on the aircraft's efficiency. This overheating also affects the aircraft's engines, increasing the likelihood of failure even for the engine's non-moving parts.

The "scramjet," while similar to the ramjet, offers a significant difference. While the incoming air is slowed down to subsonic speeds inside the ramjet, it maintains supersonic speeds through the scramjet. This means less of the kinetic energy of the fast-moving molecules is converted to thermal energy, in this case, useless heat. So the engine actually heats up less and more of the kinetic energy is used to propel the aircraft.

Will the scramjet work? No one knows for certain yet. The X-30 experimental aircraft now on the drawing boards will give engineers a chance to test the theory, but many complications and problems still have to be solved before the National Aerospace Plane

The ESA Hermes space shuttle under development

gets off the ground. Not the least of these is the question of practicality. Scramjets only begin to work at high speeds near Mach 4 but they wouldn't be useful for long on a space-bound plane since the aircraft would still have to switch over to rocket power as it nears orbit, where there is no air for the scramjet combustion. The exact speed at which these "switch-overs" happen, and whether such a speed can be economically achieved may determine the future for the scramjet and the National Aerospace Plane.

The European and Japanese Direction

The European Space Agency is also working on achieving more economical and efficient access to orbit. Still in the earliest conception, these advanced spacecraft are imagined in a variety of technologies and configurations. The French have proposed a Space Shuttle called Hermes that will give ESA space shuttle capabilities comparable, but more limited in payload capacity to the space shuttle, but already the Europeans are looking beyond to more advanced designs. West Germany's proposed "Sanger" concept would "piggyback" an upper-stage spaceplane to a lower-stage aircraft from which it would separate at high altitudes. Both aircraft would then be able to return and land at the same conventional airports.

The United Kingdom, meanwhile, struggling to find if it has a place in the future of space exploration, has proposed the "Hotol" concept. The Hotol, like the U.S. National Aerospace Plane, is a single-stage Earth-to-orbit reusable aerospace plane that could land and take off from conventional airport runways. Like the U.S. NASP, "Hotol" faces some tough engineering problems inherent in the scramjet concept, which

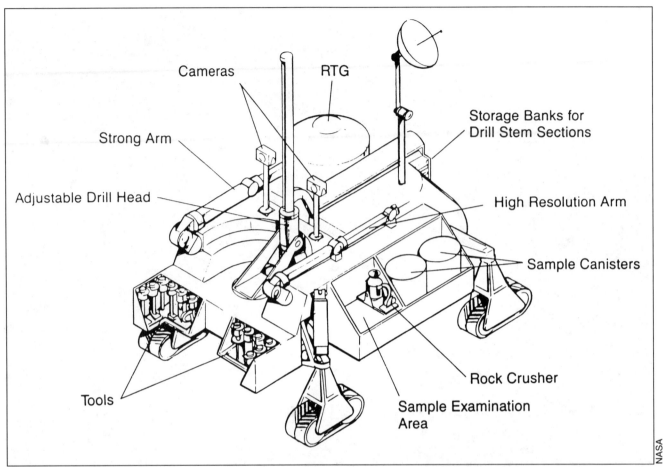

Cameras

RTG

Strong Arm

Storage Banks for
Drill Stem Sections

Adjustable Drill Head

High Resolution Arm

Sample Canisters

Tools

Rock Crusher

Sample Examination
Area

NASA

A future mission to Mars may send a rover to sample the surface prior to a manned mission

British scientists and engineers hope to solve with daring and advanced technology.

The Japanese are currently researching three different spaceplane/shuttle programs. One, like ESA's Hermes, would be small and would ride an expendable rocket into orbit. A second, called HIMES, is a highly maneuverable spaceplane, while a third would use scramjet technology.

Beyond Tomorrow

Will the scramjet technologies carry us into a new age of living and working in space? Only the future will tell. While the United States, ESA and other space-faring nations are beginning to prepare for tomorrow, a small handful of space visionaries are looking beyond tomorrow. To conquer not only near space but also deep space, as well, to push outward into the stars, is the dream of these visionaries. To achieve this dream, they see space transportation systems many generations away from those of today or tomorrow. These ideas range from the ruggedly practical (advanced new expendable launch vehicles—ELVs—

which use dramatically improved fuel combustion technologies), to the poetic—but still feasible—(solar-sailing craft that use the particles streaming out from the sun to propel them much like conventional sailing craft), to the wishful (faster-than-light travel using such science-fiction devices as "time and space warps," "anti-gravity" devices and "anti-matter drives").

Impossible? Who is to say. Certainly the basic laws of physics prohibit the realization of many such concepts and indicate the direction such dreams must follow if they are to come true. But the idea that humankind might roam freely in the Earth's skies was once a childish fantasy too, and people once called the notion that human feet might ever touch the Moon an irresponsible fantasy.

Inhabiting the Moon and Mars

If everything goes well, space stations and improved space transportation systems will bring near-space even more into our daily lives. Workers, scientists and travelers commuting regularly to and from space will

NASA

Mars Rover collecting samples

seem no different to us than today's busy business people flying across continents. The bottom line, as business school graduates like to say, is profit, and profit lies waiting in orbit just beyond the Earth's atmosphere.

Will humans ever again set foot on the Moon, or journey across space to step for the first time on the face of the red planet Mars? Does the future of humankind stretch far beyond near Earth orbit? Or, will humankind reach out no further, leaving the rest of space to be studied and explored by "robot representatives" such as the Pioneer, Viking and Voyager spacecraft? Scientists and enthusiasts are becoming increasingly divided about the future direction of America's space plans.

A large and vocal group in the space science community is uneasy about the large amounts of money and effort that have gone into "manned" space programs. These scientists, including such distinguished space science pioneers as James Van Allen, discoverer of the Van Allen belts around Earth, believe that science should come first and that space science can best be done with unmanned spacecraft.

For many others, though, advanced spacecraft and space stations are just the necessary first steps in a much broader picture. Like visionaries of the past, they see us pushing ever outward, with a return to the Moon the next logical step toward this "in person" exploration and exploitation of the space frontier.

The back-to-the-Moon segment of the space community has marshalled some pretty strong arguments. They point out that if humankind is to journey to other planets of the solar system, a Moon base would be a good testing ground to prepare for longer duration human space ventures. Furthermore, the Moon itself is rich in resources and has a much lower gravity than the Earth. These resources could be used to build larger and more complex space stations that could be lifted from the Moon much more economically than from Earth.

"Space colony" advocate and physicist Gerard K. O'Neill has proposed an ingenious device called a "mass driver" which would use long "tracks" and magnetic levitation to literally sling bucketfuls of moon materials into space to be "caught" and used for the construction of his large-scale space settlements.

The return-to-the-Moon advocates also point out the obvious advantages of an astronomical observatory on the Moon, particularly on the far side, and the opportunities to study our nearest neighbor much more intensely than has ever been done before. By starting small, with a rotating crew of six or seven astronaut-scientists, and then using the Moon's resources to gradually build a larger and more permanent base, a year-round scientific outpost could be in place a few years after the first base camp was established.

Additionally, advocates say, the Moon would make an ideal low-gravity staging point for such missions as an eventual human expedition to Mars.

They point out that humans can make decisions that robots cannot and that the natural curiosity of the human mind will explore much more creatively and flexibly than any robot we've yet built. As a result, a human expedition to the red planet could, in the long run, be the more economical and scientifically productive plan.

Destination: Mars

Ironically, advocates of a human expedition to Mars are themselves split about the merits of returning to the Moon first. While some concede the logical merits in the plan, others are afraid that too much time and energy would end up being expended on a Moon base. With a kind of "we've already been there, let's get on to something more interesting" attitude, this group is pushing for a human expedition to Mars itself as soon as possible. The "Mars now" people believe that such an ambitious project would not only mark another giant step forward for humankind, but also would excite public interest and galvanize what they see as the lethargy of the present American space program.

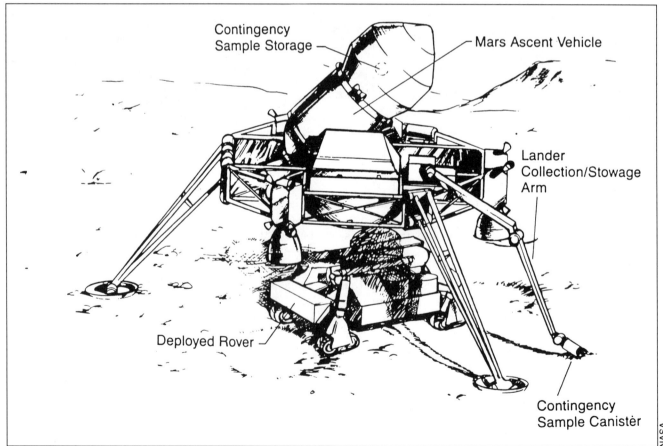

Mars Lander deploying a rover to collect samples.

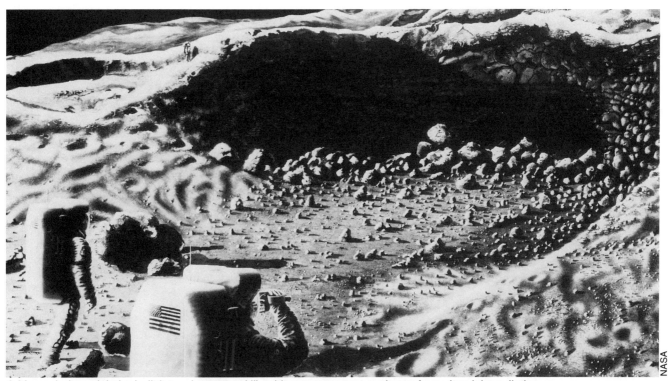

A Moon shelter might be built into a lava tunnel like this one to protect explorers from ultraviolet radiation

Future missions to the Moon or Mars will take advantage of local resources, such as elements and minerals found in ore

Will humankind return to the Moon and push on to Mars and beyond?

The decision and the responsibility are ours alone.

Will we have the courage and imagination to make that decision?

As Frederick Langbridge wrote in 1896:

"Two men look out through the same bars:
One sees the mud, and one the stars."

APPENDIX
MEN AND WOMEN IN SPACE—SOVIET MISSIONS

Mission	Cosmonauts	Dates
Vostok 1	Gagarin	April 12, 1961
Vostok 2	G. Titov	Aug. 6-7, 1961
Vostok 3	Nikolayev	Aug. 11-15, 1962
Vostok 4	Popovich	Aug. 12-15, 1962
Vostok 5	Bykovsky	June 14-19, 1963
Vostok 6	Tereshkova	June 16-19, 1963
Voskhod 1	Komarov, Yegorov, Feoktistov	Oct. 12-13, 1964
Voskhod 2	Belyayev, Leonov	Mar. 18-19, 1965
Soyuz 1	Komarov (died on reentry)	April 23-24, 1967
Soyuz 3	Beregovoy	Oct. 26-30, 1968
Soyuz 4	Shatalov	Jan. 14-17, 1969
Soyuz 5	Volynov, Yeliseyev, Khrunov	Jan. 15-18, 1969
Soyuz 6	Shonin, Kubasov	Oct. 11-16, 1969
Soyuz 7	Filipchenko, V. Volkov, Gorbatko	Oct. 12-17, 1969
Soyuz 8	Shatalov, Yeliseyev	Oct. 13-18, 1969
Soyuz 9	Nikolayev, Sevastyanov	June 1-19, 1970
Soyuz 10	Shatalov, Yeliseyev, Rukavishnikov	April 23-25, 1971
Soyuz 11	Dobrovolsky, V. Volkov, Patsayev (crew died during reentry)	June 6-30, 1971
Soyuz 12	Lazarev, Makarov	Sept. 27-29, 1973
Soyuz 13	Klimuk, Artyukhin	Dec. 18-26, 1973
Soyuz 14	Popovich, Artyukhin	July 3-19, 1974
Soyuz 15	Sarafanov, Demin	Aug. 26-28, 1974
Soyuz 16	Filipchenko, Rukavishnikov	Dec. 2-8, 1974
Soyuz 17	Gubarev, Grechko	Jan. 11-Feb. 9, 1975
Soyuz 18-1	Lazarev, Makarov (failed launch)	April 5, 1975
Soyuz 18	Klimuk, Sevastyanov	May 24-July 26, 1975
Soyuz 19	Leonov, Kubasov (Apollo-Soyuz Test Project)	July 15-21, 1975
Soyuz 21	Volynov, Zholobov	July 6-Aug. 24, 1976
Soyuz 22	Bykovsky, Aksenov	Sept. 15-23, 1976
Soyuz 23	Zudov, Rozhdestvensky	Oct. 14-16, 1976
Soyuz 24	Gorbatko, Glazkov	Feb. 7-25, 1977
Soyuz 25	Kovalenok, Ryumin	Oct. 9-11, 1977
Soyuz 26	Romanenko, Grechko	Dec. 10-Mar. 16, 1977
Soyuz 27	Dzhanibekov, Makarov	Jan. 10-16, 1978
Soyuz 28	Gubarev, Remek (Czechoslovakia)	Mar. 2-10, 1978
Soyuz 29	Kovalenok, Ivanchenkov	Jun 15-Nov. 2, 1978

Mission	Cosmonauts	Dates
Soyuz 30	Klimuk, Hermaszewski (Poland)	June 27-July 5, 1978
Soyuz 31	Bykovsky, Jaehn (German Democratic Republic)	Aug. 26-Sept. 3, 1978
Soyuz 32	Lyakhov, Ryumin	Feb. 25-Aug. 19, 1979
Soyuz 33	Rukavishnikov, Ivanov (Bulgaria)	April 10-12, 1979
Soyuz 35	Popov, Ryumin	April 9-Oct. 11, 1980
Soyuz 36	Kubasov, Farkas (Hungary)	May 26-June 3, 1980
Soyuz T-2	Malyshev, Aksenov	June 5-9, 1980
Soyuz 37	Gorbatko, Tuan (Vietnam)	July 23-31, 1980
Soyuz 38	Romanenko, Tamayo-Mendez (Cuba)	Sept. 18-26, 1980
Soyuz T-3	Kizim, Makarov, Strekalov	Nov. 27-Dec. 10, 1980
Soyuz T-4	Kovalenok, Savinykh	Mar. 12-May 26, 1981
Soyuz 39	Dzhanibekov, Gurragcha (Mongolia)	Mar. 22-30, 1981
Soyuz 40	Popov, Prunariu (Romania)	May 14-22, 1981
Soyuz T-5	Berezovoy, Lebedev	May 13-Dec. 10, 1982
Soyuz T-6	Dzhanibekov, Ivenchenkov, Chrétien (France)	June 24-July 2, 1982
Soyuz T-7	Popov, Serebrov, Savitskaya	Aug. 19-27, 1982
Soyuz T-8	V. Titov, Strekalov, Serebrov	April 20-22, 1983
Soyuz T-9	Lyakhov, Alexandrov	June 27-Nov. 23, 1983
Soyuz T-10-1	V. Titov, Strekalov (launch pad abort)	Sept. 26, 1983
Soyuz T-10	Kizim, V. Solovyov, Atkov	Feb. 8-Oct. 2, 1984
Soyuz T-11	Malyshev, Strekalov, Sharma (India)	April 3-11, 1984
Soyuz T-12	Dzhanibekov, Savitskaya, Volk	July 17-29, 1984
Soyuz T-13	Dzhanibekov, Savinykh	June 6-Sept. 26, 1985
Soyuz T-14	Vasyutin, Grechko, A. Volkov	Sept. 17-Nov. 21, 1985
Soyuz T-15	Kizim, V. Solovyov	Mar. 13-July 16, 1986
Soyuz TM-2	Romanenko, Laveikin	Feb. 5-Dec. 28, 1987
Soyuz TM-3	Viktorenko, Alexandrov (USSR), Faris (Syria)	July 22-30, 1987
Soyuz TM-4	Levchenko, V. Titov, Manarov	Dec. 21, 1987-Dec. 21, 1988
Soyuz TM-5	A. Solovyov, Savinikh, Alexandrov (Bulgaria)	June 7-15, 1988
Soyuz TM-6	Lyakhov, Polyakov, Mohmand (Afghanistan)	Aug. 29-Sept. 7, 1988
Soyuz TM-7	A. Volkov, Krikalev, Chrétien (France)	Nov. 26-Dec. 21, 1988

MEN AND WOMEN IN SPACE—U.S. MISSIONS

Mission	Astronaut(s)	Dates
Mercury-Redstone 3	Shepard	May 5, 1961
Mercury-Redstone 4	Grissom	July 21, 1961
Mercury-Atlas 6	Glenn	Feb. 20, 1962
Mercury-Atlas 7	Carpenter	May 24, 1962
Mercury-Atlas 8	Schirra	Oct. 3, 1962
Mercury-Atlas 9	Cooper	May 15-16, 1963
Gemini-Titan 3	Grissom, Young	Mar. 23, 1965
Gemini-Titan 4	McDivitt, White	June 3-7, 1965
Gemini-Titan 5	Cooper, Conrad	Aug. 21-29, 1965
Gemini-Titan 7	Borman, Lovell	Dec. 4-18, 1965
Gemini-Titan 6-A	Schirra, Stafford	Dec. 15-16, 1965
Gemini-Titan 8	Armstrong, Scott	Mar. 16, 1966
Gemini-Titan 9-A	Stafford, Cernan	June 3-6, 1966
Gemini-Titan 10	Young, Collins	July 18-21, 1966
Gemini-Titan 11	Conrad, Gordon	Sept. 12-15, 1966
Gemini-Titan 12	Lovell, Aldrin	Nov. 11-15, 1966
Apollo 1	Grissom, White, Chaffee (crew killed in launchpad fire)	Jan. 27, 1967
Apollo-Saturn 7	Schirra, Eisele, Cunningham	Oct. 11-22, 1968
Apollo-Saturn 8	Borman, Lovell, Anders	Dec. 21-27, 1968
Apollo-Saturn 9	McDivitt, Scott, Schweickart	Mar. 3-13, 1969
Apollo-Saturn 10	Stafford, Young, Cernan	May 18-26, 1969
Apollo-Saturn 11	Armstrong, Collins, Aldrin	July 16-24, 1969
Apollo-Saturn 12	Conrad, Gordon, Bean	Nov. 14-24, 1969
Apollo-Saturn 13	Lovell, Swigert, Haise	April 11-17, 1970
Apollo-Saturn 14	Shepard, Roosa, Mitchell	Jan. 31-Feb. 9, 1971
Apollo-Saturn 15	Scott, Worden, Irwin	July 26-Aug. 7, 1971
Apollo-Saturn 16	Young, Mattingly, Duke	April 16-27, 1972
Apollo-Saturn 17	Cernan, Evans, Schmitt	Dec. 7-19, 1972
Skylab SL-2	Conrad, Kerwin, Weitz	May 25-June 22, 1973
Skylab SL-3	Bean, Garriott, Lousma	July 28-Sept. 25, 1973
Skylab SL-4	Carr, Gibson, Pogue	Nov. 15-Feb. 8, 1974
Apollo-Soyuz Test Project (Apollo 18)	Stafford, Brand, Slayton	July 15-24, 1975
STS-1 U.S. Space Shuttle Columbia	Young, Crippen	April 12-14, 1981
STS-2 U.S. Space Shuttle Columbia	Engle, Truly	Nov. 12-14, 1981
STS-3 U.S. Space Shuttle Columbia	Lousma, Fullerton	Mar. 22-30, 1982
STS-4 U.S. Space Shuttle Columbia	Mattingly, Hartsfield	June 27-July 4, 1982
STS-5 U.S. Space Shuttle Columbia	Brand, Overmyer, Allen, Lenoir	Nov. 11-16, 1982
STS-6 U.S. Space Shuttle Challenger	Weitz, Bobko, Peterson, Musgrave	April 4-9, 1983
STS-7 U.S. Space Shuttle Challenger	Crippen, Hauck, Ride, Fabian, Thagard	June 18-24, 1983
STS-8 U.S. Space Shuttle Challenger	Truly, Brandenstein, D. Gardner, Bluford, W. Thornton	Aug. 30-Sept. 5, 1983

Mission	Astronaut(s)	Dates
STS-9 U.S. Space Shuttle Columbia—Spacelab 1	Young, Shaw, Garriott, Parker, Lichtenberg, Merbold (ESA)	Nov. 28-Dec. 8, 1983
41-B U.S. Space Shuttle Challenger	Brand, Gibson, McCandless, Mc-Nair, Stewart	Feb. 3-11, 1984
41-C U.S. Space Shuttle Challenger	Crippen, Scobee, Van Hoften, G.Nelson, Hart	April 6-13, 1984
41-D U.S. Space Shuttle Discovery	Hartsfield, Coats, Resnik, Hawley, Mullane, C. Walker	Aug. 30-Sept. 5, 1984
41-G U.S. Space Shuttle Challenger	Crippen, McBride, Ride, Sullivan, Leestma, Garneau (Canada), Scully-Power	Oct. 5-Oct. 13, 1984
51-A U.S. Space Shuttle Discovery	Hauck, D. Walker, D. Gardner, A Fisher, Allen	Nov. 8-16, 1984
51-C U.S. Space Shuttle Discovery	Mattingly, Shriver, Onizuka, Buchli, Payton	Jan. 24-27, 1985
51-D U.S. Space Shuttle Discovery	Bobko, Williams, Seddon, Hoffman, Griggs, C. Walker, Garn	April 12-19, 1985
51-B U.S. Space Shuttle Challenger—Spacelab 3	Overmyer, Gregory, Lind, Thagard, W. Thornton, van den Berg, Wang	April 29-May 6, 1985
51-G U.S. Space Shuttle Discovery	Brandenstein, Creighton, Lucid, Fabian, Nagel, Baudry (France), Al-Saud (Saudi Arabia)	June 17-24, 1985
51-F U.S. Space Shuttle Challenger—Spacelab 2	Fullerton, Bridges, Musgrave, England, Henize, Acton, Bartoe	July 29-Aug. 6, 1985
51-I U.S. Space Shuttle Discovery	Engle, Covey, van Hoften, Lounge, W. Fisher	Aug. 27-Sept. 3, 1985
51-J U.S. Space Shuttle Atlantis	Bobko, Grabe, Hilmers, Stewart, Pailes	Oct. 3-7, 1985
61-A U.S. Space Shuttle Challenger—Spacelab D-1	Hartsfield, Nagel, Buchli, Bluford, Dunbar, Furrer (West Germany), Messerschmid (West Germany), Ockels (ESA)	Oct. 30-Nov. 6, 1985
61-B U.S. Space Shuttle Atlantis	Shaw, O'Connor, Cleave, Spring, Ross, Neri-Vela (Mexico), C. Walker	Nov. 26-Dec. 3, 1985
61-C U.S. Space Shuttle Columbia	Gibson, Bolden, Chang-Diaz, Hawley, G. Nelson, Cenker, B. Nelson	Jan. 12-18, 1986
51-L U.S. Space Shuttle Challenger	Scobee, Smith, Resnik, Onizuka, McNair, Jarvis, Mc-Auliffe (crew died during launch)	Jan. 28, 1986
STS-26 U.S. Space Shuttle Discovery	Hauck, Covey, G. Nelson, Hilmers, Lounge	Sept. 29-Oct. 3, 1988
STS-27 U.S. Space Shuttle Atlantis	Gibson, Gardner, Ross, Mullane, Shepherd	Dec. 2-6, 1988
STS-28 U.S. Space Shuttle Discovery	Coats, Blaha, Buchli, Springer, Bagian	Mar. 13-18, 1989

GLOSSARY

abort To end a mission or activity prematurely (before planned completion).

airlock An intermediate chamber between the airlessness of space and the interior pressurized quarters of a spacecraft. In space, people typically enter and exit a spacecraft by passing through an airlock; otherwise, the spacecraft would depressurize and the atmosphere would escape through the open hatch.

aerodynamics The branch of science that studies the behavior and flow of air around objects, and the forces it exerts on those objects, including resistance or drag, pressure, etc.

altimeter An instrument that measures altitude.

Apollo The NASA project (and the name of the spacecraft) that landed 12 men on the Moon between 1969 and 1972.

artificial intelligence The effort to duplicate human intelligence using computers.

ASTP Apollo-Soyuz Test Project, the historic cooperative program between the United States and the USSR resulting in a joint space mission in July 1975. Two Soviets and three Americans met in space when the orbiting Apollo and Soyuz spacecraft linked up. The term "test project" was used because only one mission was planned.

ATM Apollo Telescope Mount—Skylab's solar observatory module, the first manned astronomical observatory designed for solar research from Earth orbit.

ballistic descent A flight path that returns a spacecraft to Earth in the same manner as a bullet or other falling object.

ballistic trajectory The curved path of a projectile, such as a ballistic missile or a bullet, after the thrust or propelling force has ended. Because early spacecraft, such as the early versions of Soyuz, had very little control, they had to be aligned exactly right at launch in order to enter orbit properly. They traveled along ballistic trajectories.

booster In a multistage rocket, any of the rockets that provide the early stages of propulsion, including the initial stage that provides power for the launching and initial part of the flight.

byte One "word" of computer data.

Cape Canaveral Location of NASA Kennedy Space Center and the Air Force station from which all Mercury and Gemini missions were launched as well as, to this day, most unmanned missions. Called Cape Kennedy between 1963 and 1973.

cargo bay The area of a spacecraft (most commonly the U.S. Shuttle) used to transport equipment, experiments and so on. Satellites to be launched during a Shuttle mission, for example, are carried in the cargo bay, located behind the crew quarters.

Chibis vacuum suit Special low-pressure trousers worn by Soviet cosmonauts during their work day aboard Salyut 6 to pull blood from their upper bodies into their legs and keep the heart from getting lazy in the physically undemanding environment of weightlessness.

CM The Command Module, one of the three main parts of the Apollo spacecraft. Shaped like a cone, it contained the crew quarters, and was the only portion of the spacecraft that returned to Earth at the completion of a mission.

coma The cloud of dust and gas surrounding the nucleus (or center) of a comet.

CSM The Command/Service Module (Command Module and Service Module combined) of the Apollo spacecraft.

decaying orbit When an orbiting satellite or spacecraft gradually moves out of its orbit and eventually falls, usually due to slowing velocity casued by atmospheric drag, the orbit is said to decay or deteriorate.

deep space The outer Solar System and beyond.

DM Docking module, a section of a spacecraft (such as Skylab) that enables docking or joining with another spacecraft.

docking When two spacecraft connect or "link up" with each other while in orbit.

EDT Eastern Daylight Time (one hour ahead of EST), the official time used for all launches made during summer months from Kennedy Space Center in Florida.

electromagnetic radiation Waves or rays created by variations in magnetic and electric fields—including radio waves, infrared, visible light, ultraviolet, X-rays and gamma rays.

electromagnetic spectrum The entire range of electromagnetic radiation from gamma rays (the shortest wavelength) to radio waves (the longest) and including visible light.

EST Eastern Standard Time, the official time used for all launches made during the winter months from Kennedy Space Center (located in Florida).

EVA Extravehicular activity—a space walk (outside the protection of a spacecraft).

flight trajectory The path traveled by a flying object, such as a rocket or spacecraft.

fusion The process of combining two substances by melting.

gamma ray telescope A special telescope used by astronomers to detect gamma rays emitted by objects in space (such as galaxies and quasars). Gamma rays, which are extremely short-wavelength electromagnetic radiation, cannot penetrate the Earth's atmosphere, so they can be observed only from space (for example, from a satellite or spacecraft).

Gemini NASA's second manned space project, lasting from 1965-1966. The Gemini spacecraft carried two astronauts on each mission.

geostationary orbit A special circular *geosynchronous orbit* at the Equator. From the Earth a satellite in geostationary orbit seems to hover motionless in one spot above the Earth although it is actually traveling as fast as the Earth is rotating, one revolution every 24 hours.

geosynchronous orbit A special orbit 22,300 miles above the Earth, where a satellite's movement is synchronized (-*synchronous*) with the Earth (*geo*-).

Getaway Special One of a series of low-cost, self-contained scientific experiments developed by industry, schools, individuals, and so on, carried aboard the Space Shuttle in a special payload canister.

gigahertz One billion cycles a second.

high-resolution imaging The use of highly sensitive instruments (such as a camera) to produce close-up images. Cameras aboard certain military satellites, for example, can reportedly photograph areas on the Earth's surface just a few inches across.

infrared Radiation lying just outside the visible spectrum at the red end, with wavelengths shorter than radio waves. Infrared radiation cannot be seen with the naked eye, but infrared radiation from many stars and galaxies can be detected by an infrared telescope, such as the one carried aboard IRAS (Infrared Astronomical Satellite) in 1983.

Intelsat The International Telecommunications Satellite Consortium, a commercial group responsible for financing many telecommunications satellites.

JSC Johnson Space Center, the NASA Center at Clear Lake, Texas, near Houston. Previously known as the Manned Spacecraft Center, this site houses NASA Mission Control, which manages all manned spaceflights after lift-off from Cape Canaveral Air Force Station (Mercury, Gemini, and most unmanned missions) or from Kennedy Space Center (Apollo missions and, later, Skylab and Shuttle missions).

keeping station Maneuvering a spacecraft so that it remains near another object in orbit.

KSC Kennedy Space Center, the NASA center on Merritt Island in Florida. All Apollo missions were launched from KSC (as well as, later, all Skylab and Shuttle missions).

LANDSAT Name given to a series of Earth observation satellites.

lift-off Ascent of a spacecraft from the launch pad (to at least two inches).

materials processing Manufacturing or transforming substances, an activity of interest in space, since, for example, near-perfect ball bearings and crystals can be created in the weightless environment of space.

LM The Lunar Module, one of the three main parts of the Apollo spacecraft, used to land astronauts on the Moon.

megahertz A million cycles per second.

Mercury The NASA project that included the first U.S. manned missions into space.

Mercury 7 The first seven U.S. astronauts chosen to participate in the Mercury missions.

metric ton 1.1 U.S. tons or 2,200 pounds.

microwave Very high frequency radio waves.

Mir The "new-generation" Soviet space station launched February 20, 1986. The name means "peace."

mission specialist An astronaut on a Space Shuttle crew responsible for some aspect of mission operations.

MMU Manned maneuvering unit, a special jet-propelled backpack that enables Shuttle astronauts to work without a tether to the spacecraft during EVA. First used by Bruce McCandless and Robert Stewart during STS-41B in 1984, when McCandless flew the distance of half a football field from the spacecraft.

multispectral scanner An instrument that measures several regions of the electromagnetic spectrum.

multistage rocket Several rockets fired in combination to achieve greater heights.

NASA National Aeronautics and Space Administration, the U.S. space agency.

NASDA National Space Development Agency of Japan.

near infrared Shorter infrared wavelengths (those closest to the red end of the visible spectrum), as opposed to far infrared, a term used to describe longer infrared wavelengths.

near space The area of space closest to the Earth, as opposed to the outer Solar System and the starry regions of "deep space" beyond.

neutral buoyancy tank A very large "swimming pool," filled with water, used for testing equipment and procedures for use in space. By using weights that correct for the tendency of objects and people to float, astronauts and technicians swimming underwater in this tank can simulate the weightlessness of space.

NOAA National Oceanic and Atmospheric Administration, an agency of the United States Department of Commerce concerned with the study of the atmosphere and oceans.

oxidizer A material that supplies oxygen so that combustion can take place in a rocket engine.

passive communications satellite A satellite (such as the Echo series) that reflects signals from its surface without originating any signals.

payload Anything that's not part of the functioning of a rocket or spacecraft but is transported by it to carry out a purpose or mission; cargo.

payload specialist A non-astronaut who goes into space aboard the Space Shuttle to perform specific scientific experiments or other work.

PDT Pacific Daylight Time (one hour ahead of standard time), the official time used for all Shuttle landings made during the summer months at Edwards Air Force Base in California.

projectile Any object propelled forward by a force. For example, a bullet or a spacecraft.

propulsion A force that propels, or pushes, an object forward.

Proton launcher The Soviet "D-class" launch vehicle, designed to lift heavy loads such as the Salyut and Mir space stations and the first Soviet booster not developed directly from a military rocket. Introduced in July 1965, it is an enlarged version of the smaller "A-class" booster (used to launch all three Sputnik satellites, as well as Vostok, Voskhod, Soyuz and Progress missions). The USSR is currently promoting the Proton launcher for use by the rest of the world as a commercial launch vehicle.

PST Pacific Standard Time, the official time used for all Shuttle landings made during the winter months at Edwards Air Force Base.

radar *R*adio *d*etecting *a*nd *r*anging. A device that bounces ultrahigh-frequency radio waves off a target object and, from changes observed in the reflected signal, determines characteristics (such as distance and direction) of the object.

radio telescope An instrument—including an antenna or system of antennas—used for measuring and recording the radio frequency signals from objects in the universe such as Jupiter, the Sun, pulsars and so on.

ramjet An engine technology used on supersonic aircraft that makes use of the aircraft's speed to gulp (or "ram") air into the engine, where it is slowed down, mixed with fuel, ignited, and expelled through the rear nozzle of the engine to achieve greater speeds.

real time The actual time in which an event or process takes place, as opposed to delayed time or projected time.

rendezvous When two spacecraft in orbit are brought together almost close enough to touch.

retrofire Firing rockets in the opposite direction form the direction of flight to slow a spacecraft (a procedure used especially during reentry).

retro-rockets Rockets used to slow the speed of a spacecraft. They are fired in the opposite direction from the direction of flight ("retro-" or backward).

Salyut The name given the first seven Soviet space stations. The first Salyut was launched April 19, 1971.

Salyut 7, launched April 19, 1982, was "retired" though still orbiting when the Mir space station was launched in 1986. The name means "salute," honoring Soviet cosmonaut Yuri Gagarin, the first human in space.

Saturn rocket A powerful, giant rocket designed to send Apollo missions to the Moon.

scramjet An aircraft engine technology similar to the *ramjet* but which does not slow the air down when it enters the engine. Instead it makes use of the air's extremely high speed to help propel the aircraft.

simulator Any device that makes it possible to practice an activity outside the usual environment and without using the actual equipment (for example, a flight simulator). For space missions, underwater simulators (in neutral buoyancy tanks) have proven highly useful to both Soviet and American manned programs. To simulate weightlessness, a mock-up spacecraft can be placed in a huge underwater tank where swimmers fitted with weights can achieve "neutral buoyancy"—they neither sink nor float to the surface. Astronauts and cosmonauts train in such simulators, and ground crews often help during missions by trying out solutions to problems in an underwater simulator.

SM The Service Module, one of the three main parts of the Apollo spacecraft. Cylindrical in shape, it contained fuel, supplies and engines.

solar array A panel (often wing-shaped) of solar cells used for power, for example, aboard Soyuz, Skylab and Mir.

solar cell A device used to convert rays from the Sun directly into electrical power.

solar flare A sudden short-lived outburst of energy from a small area of the Sun's surface.

solid fuel A fuel, such as gunpowder, that is a solid (as opposed to being in a liquid or a gaseous state).

sounding rocket A rocket used to obtain information about the atmosphere.

Soyuz A Soviet spacecraft originally designed to carry three cosmonauts. Its name, meaning "union," indicates its main mission, to provide transportation to and from a space station, where it could dock during missions aboard the station. The descendants of this spacecraft, the Soyuz TM series, are still being flown today.

spaceplane A vehicle that can take off horizontally from Earth, achieve speeds many times faster than sound, fly above the atmosphere to space and then return and land like an airplane.

space suit A suit equipped with life-support provisions (such as oxygen, temperature regulation, pressurization, protection against radiation and so on) to allow the wearer to function in space outside the spacecraft, an EVA suit.

speed of sound At 32 degrees F (0 degrees C) the speed of sound in air is about 760 mph (332 meters per second).

Sputnik The first artificial satellite, launched by the USSR on October 4, 1957. The name means, roughly, "traveling companion," and two more were also later launched.

suborbital Describes a flight that does not make an orbit around the Earth, such as Alan Shepard's Mercury flight.

subsonic Slower than the speed of sound.

supersonic Faster than the speed of sound.

thrust The push forward caused in reaction to a high-speed jet of fluid or gases discharged in the opposite direction from a rocket's nozzle.

thrusters Rocket engines, especially those used for maneuvering a spacecraft.

transfer module An airlock through which one can enter or leave a spacecraft.

Tyuratam The Soviet launch site, located on the broad, flat steppes of Central Asia, about 200 miles from the town of Baikonur (the name often used by the Soviets for the launch site). Sometimes also referred to as "Starry Town."

UHF Ultrahigh frequency, a radio frequency in the extremely high range (between 3 GHZ [gigahertz] and 0.3 GHz—compared to VHF, very high frequency, which is between 300 MHz [megahertz] and 30 MHz).

USGS United States Geological Survey, an agency of the U.S. Department of the Interior concerned with the study of geology.

ultraviolet Radiation having a wavelength shorter than visible light (just beyond the violet end of the visible spectrum and therefore can't be seen by the naked eye) and longer than X-rays.

upper stage In a multistage rocket, a booster rocket that takes over after the first-stage rocket has burned its fuel.

venting out Expelling a gas.

Voskhod A modified Soviet Vostok spacecraft, used for two manned missions in 1964-65. Its name means "Ascent."

Vostok The first manned Soviet spacecraft. Its name means "East."

wind shear Aerodynamic stress caused by changes in wind direction.

window of opportunity The space of time during which an action can be taken successfully, such as a window of opportunity for a launch or for a descent from orbit.

X-ray telescope An instrument used to collect and measure X-ray radiation from objects in space. Many nearby stars and other astronomical objects emit X-rays, but one example is the well-known Crab nebula, left over from an ancient supernova.

SUGGESTIONS FOR FURTHER READING

Books:

Associated Press. *Moments in Space.* New York: Gallery Books, 1986.

Baker, David. *The History of Manned Space Flight,* Revised Edition. New York: Crown Publishers, Inc., 1982.

————. *The Rocket: The History and Development of Rocket & Missile Technology.* New York: Crown Publishers, Inc., 1978.

Belew, Leland F. *Skylab, Our First Space Station.* Washington, DC: NASA, 1977.

Benford, Timothy B., and Brian Wilkes. *The Space Program Quiz & Fact Book.* New York: Harper & Row, 1985.

Bond, Peter. *Heroes in Space: From Gagarin to Challenger.* New York: Basil Blackwell, Inc., 1987.

Caprara, Giovanni. *The Complete Encyclopedia of Space Satellites: Every Civil and Military Satellite of the World Since 1957.* English Translation. New York: Portland House, 1986.

Cassutt, Michael. *Who's Who in Space: The First 25 Years.* Boston: G.K. Hall & Co., 1987.

Clark, Phillip. *The Soviet Manned Space Program: An Illustrated History of the Men, the Missions, and the Spacecraft.* New York: Orion Books, 1988.

Collins, Michael. *Liftoff: The Story of America's Adventure in Space.* New York: Grove Press, 1988.

Cooper, Henry S.F. Jr. *A House in Space.* New York: Holt, Rinehart and Winston, 1976.

Dewaard, E. John and Nancy. *History of NASA: America's Voyage to the Stars.* New York: Exeter Books, 1984.

Gatland, Kenneth, et al. *The Illustrated Encyclopedia of Space Technology: A Comprehensive History of Space Exploration.* New York: Harmony Books, 1981.

Gurney, Gene, and Jeff Forte. *Space Shuttle Log: The First 25 Flights.* Foreword by James M. Beggs, former NASA Administrator and Preface by Milton A. Silveira, former NASA Chief Engineer. Blue Ridge Summit, PA: Aero (A Division of Tab Books), 1988.

Hart, Douglas. *The Encyclopedia of Soviet Spacecraft.* New York: Exeter Books, 1987.

Hurt, Harry, III. *For All Mankind.* New York: Atlantic Monthly Press, 1988.

Johnson, Nicholas L. *Handbook of Soviet Manned Space Flight.* San Diego: Univelt, Inc., 1980.

Kerrod, Robin. *The Illustrated History of NASA.* New York: Gallery Books, 1986.

McAleer, Neil. *The Omni Space Almanac.* New York: World Almanac, 1987.

McDonough, Thomas R. *Space: The Next Twenty-Five Years.* New York: John Wiley, 1987.

McDougall, Walter A. *...the Heavens and the Earth: A Political History of the Space Age.* New York: Basic Books, Inc., 1985.

Oberg, James E. *Red Star in Orbit: The Inside Story of Soviet Failures and Triumphs in Space.* New York: Random House, 1981.

————, and Alcestis R. Oberg. *Pioneering Space: Living on the Next Frontier.* Foreword by Isaac Asimov. New York: McGraw-Hill, 1986.

Pioneering the Space Frontier: The Report of the National Commission on Space. (Paine Commission Report.) New York: Bantam, 1986.

Report to the President by the Presidential Commission on the Space Shuttle Challenger Accident. (Rogers Commission Report.) Washington, DC, 1986.

Silvestri, Goffredo, et al. *Quest for Space: Man's Greatest Adventure—the Facts, the Machines, the Technology.* Trans. Simon Pleasance. New York: Crescent Books, 1987.

Smolders, Peter. *Soviets in Space.* Trans. Marian Powell. New York: Taplinger Publishing Co., Inc., 1974.

Von Braun, Wernher, et al. *Space Travel: A History.* An update (and Fourth Edition) of *History of Rocketry & Space Travel.* New York: Harper & Row, 1985.

Yenne, Bill. *The Astronauts: The First 25 Years of Manned Space Flight.* New York: Exeter Books, 1986.

————, *The Encyclopedia of U.S. Spacecraft.* New York: Exeter Books, 1985.

Periodicals:

Ad Astra. Washington, DC: National Space Society.

Air & Space. Washington, DC: Smithsonian Institution.

Final Frontier. Minneapolis, MN: Final Frontier Publishing Co.

Spaceflight. London: British Interplanetary Society.

INDEX